# THE GEOGRAPHY COLORING BOOK

# THE GEOGRAPHY COLORING BOOK

## WYNN KAPIT

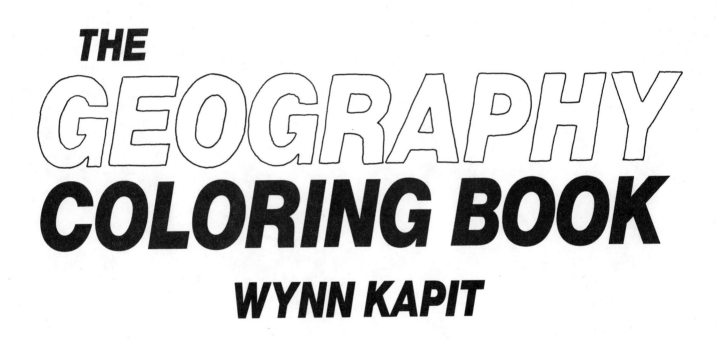

HarperCollins*Publishers*

# ACKNOWLEDGMENTS

I wish to thank Kathryn Graehl for her fine work as copy editor on this book.
Her general input and attention to detail were invaluable.

I also want to thank the staff of The TypeStudio for their excellent
job of typesetting. I want to give special thanks to Jill Breedon
whose contributions went well beyond her job as proofreader.


Wynn Kapit

Sponsoring Editor: Bonnie Roesch

Cover Printer: New England Book Components

**The Geography Coloring Book**

93   9 8 7 6 5 4

# CONTENTS

# CONTENTS

# CONTENTS

## OCEANIA

## AFRICA

## POLAR REGIONS

## WORLD THEMATIC MAPS

# PREFACE

Take your pens or pencils and color your way across the United States and around the world. With the aid of this unique and enjoyable educational tool you should be able to accomplish the following: recognize countries by shape as well as location; gain a sense of the relative sizes of nations and states; visualize the location of a nation in the context of its continent; know the names of the capitals and major cities; and identify and locate important rivers, mountain ranges, and bodies of water. And finally, you will learn essential as well as interesting facts about each of the states or nations that you have colored.

Included in each brief description are facts concerning size, population, capital city, type of government, official languages, predominant religions, major exports, and general climate. In addition, there is information that deals with some unique qualities of that particular country. It might be a matter of unusual size, population distribution, economic productivity, geographic formations and landmarks, cultural facts, famous residents, or matters of historic, military, or political significance. I have tried, wherever possible, to relate countries to the events of the day so you will have a greater insight and appreciation of the what, where, and why of the news.

This book is intended to serve as an enjoyable introduction to the countries of the world and the states of the United States. The maps have been simplified in outline and detail in order to facilitate coloring and recognition. The circles (squares for the capitals) which represent the major cites provide only a general idea of their actual shapes and sizes. The reader should keep in mind that there are hundreds of cities and towns in each country that have not been included. Similarly, only the major rivers are shown (the direction of their flow is indicated by tiny arrows). The islands that are included are usually a small fraction of the actual number.

In selecting the views for the major and supplemental maps, I have tried to give the reader a global conception, something that is often unrealized in the study of flat maps. I have avoided the use of Mercator projections, which greatly distort sizes in the northern and southern latitudes—Greenland appears larger than Africa, for example. A limited number of horizontal lines of latitude have been included to indicate relative distances from the equator or from the north and south poles. An awareness of such distances is usually helpful in estimating climatic conditions. For the purpose of simplification I have eliminated most vertical lines of longitude, which are basically used to indicate east/west distances.

You may wish to color the World Thematic Maps (Plates 42–47) after you become more familiar with the individual nations: the countries are shown on the maps but are not identified. Even if you don't color those maps first, I strongly recommend that you read the brief text that accompanies Plates 42–47. The text provides an introduction and background for many of the subjects and terms that appear frequently throughout the book—matters dealing with weather and climate, wind patterns and ocean currents, varieties of vegetation and how land is used, population and race, and languages and religions.

In illustrating and writing this book, my sense of wonder about the magnificent and varied world we live in was greatly enhanced. I hope you will share my enthusiasm and become inspired to further your knowledge by consulting the many fine atlases, books, and encyclopedias that treat this material with much greater detail.

But before you plunge in, I urge you to carefully read "How to Use and Color This Book" and learn what the various symbols represent. Spending a few minutes reading through this material will enable you to get the maximum benefit and enjoyment from this book.

WYNN KAPIT

# HOW TO USE AND COLOR THIS BOOK

Please take the few minutes needed to read through these instructions and recommendations. They will enable you to derive much more from this book than you might get if you were to plunge in without guidance. In fact, it would be wise for you to refer back to this page a number of times, as it is difficult to digest all these instructions in a single reading. This information will be more significant once you have had some practice with the coloring process. Most of it is just a matter of common sense and will become quite obvious once you get started, but there are certain things to look out for and symbols to take note of.

## WHAT MATERIALS TO USE

I recommend that you use fine-pointed felt-tip pens or colored pencils. Do not use crayons. Twelve colors, including a medium gray, should give you enough variety. Some plates will require more than 12 colors. In those cases you will have to use the same color more than once. The more colors you have at your disposal, the better your results will be and the more fun you will have in doing the book. If you have access to an art supply store that sells pens or pencils separately, you will be able to buy mostly lighter colors. Light colors do not cover up the surface detail on the maps, so the end result is more pleasing. If you are limited to a conventional color selection, try to use the lightest colors on the largest countries. Dark colors used on large areas tend to dominate the page.

## HOW THIS BOOK IS ARRANGED

This book is made up of individual plates. Each plate consists of two pages—usually, a large map on the left and a text page on the right (which may have supplementary maps). The names of the countries are printed in outlined letters on the text page; color the name with the same color you use to color the country.

The book is divided into sections, each dealing with a separate continent. The first plate in each section is a political map introducing you to the countries of that continent. The next plate shows physical features of the continent: major rivers, mountains, mountain peaks, or land regions. The remaining plates of the section cover the individual countries, grouped according to region (northwestern, southcentral, etc.). On these regional plates you will generally color a country three times: (1) in outline form on the large map, (2) in a small separate drawing, and (3) in its location on a map of the continent.

Study the cover of this book to see a page that has been completely colored. Only on the cover do the names of the countries appear on the same page as the map. This was done in order to demonstrate how the names look when colored.

# HOW TO COLOR

It is best to work through this book from the beginning. If you start with a later section, please begin with the first plate of that section and work through the rest of the plates in that section in the order they appear.

*Color only areas on the map that are within the dark outlines. They are labeled with the same letter that follows the name of the country on the text page.* The area to be colored might be the outline of a country (the space between the dark border and the dotted parallel line—see book cover). In other cases the entire country should be colored, except for the square representing the capital city. Where a group of similar areas, such as islands off the coast of a country, is labeled in only one or two places, color them all with the same color.

*Use a different color for each letter.* If you run out of colors then it is all right to repeat them, but try to avoid using the same color on adjacent countries (you can accomplish this by coloring a country before you color its name on the opposite page). Sometimes you will see the same letter with different superscripts ($A^1$, $A^2$). This means that these areas may be different from each other but are still related in some way that would justify using the same color.

Take special note of two symbols. The asterisk ( ✳ ) tells you to color gray anything labeled by it: a heading for a list of names or the names of large bodies of water bordering the continents. The "do not color" sign (-¦-) tells you to leave uncolored anything labeled by it.

It is generally a good idea to color the name first and then the country. On some pages, color notes (CN) advise you how to color that particular page or what to take note of. Read these color notes before coloring. You may wish to read about a country before coloring it; you will find that the act of coloring will have more meaning if you know something about the country.

When a city (represented by a circle or square) falls within the colored outline, leave it uncolored. When it falls within the uncolored interior, color it with the same color you used on the outline (see book cover). Apply this rule to mountain peaks (triangles). Be aware that the major cities often cover a wider geographical area than the circle symbol suggests. The size of the circle or square reflects the approximate population size of the city itself (see "What the Symbols Mean" for the population sizes) but not the size of the surrounding metropolitan area, which usually has a much larger population. Likewise, the population figures in the text apply only to cities and not metropolitan areas.

Occasionally you will come across a large lake that separates the borders of two countries or states (see Lake Titicaca between Peru and Bolivia on the cover). Though in most cases the borders usually meet in the center of the lake, for purposes of showing the lake clearly on these maps the borders are drawn around the edge of the lake and you are asked to leave the lake uncolored.

Where the name of a city or river or another feature appears within one of the boundary outlines that is to be colored, it is all right to color over it with your pen or pencil. This is one of the main reasons you would want to use light colors as much as you possibly can. But even with dark pencils you can usually lighten the color by not pressing as hard.

When coloring these plates, take time to look at the surface detail of the country you are coloring. Look to see what the neighboring countries are. Are the major cities confined to certain regions? Can you figure out why? Follow the directional flow of the major rivers. Do they play a role in population distribution? Can you predict the climate of the country by its terrain, its distance from the equator or poles, or its proximity to a major body of water? And always take note of the scale of distances on each map—the scale varies from plate to plate.

# WHAT THE SYMBOLS MEAN

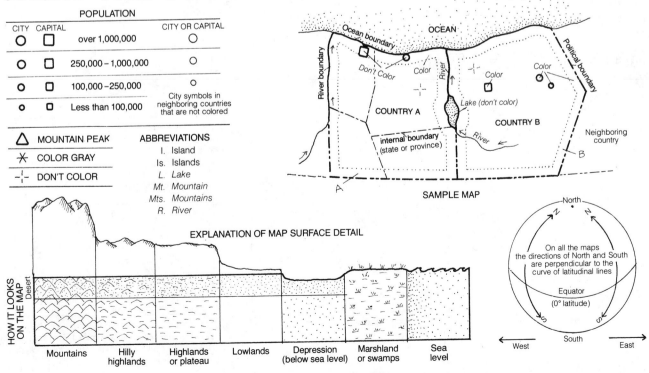

# A GLOSSARY OF GEOGRAPHICAL TERMINOLOGY

Before you begin coloring the actual plates you may wish to warm up with this introduction to geographical terminology. If you don't have the 21 colors needed to color A–U, feel free to repeat as many of them as needed.

Begin by coloring the word "archipelago," labeled "A," and use the same color on the part of the illustration below that has the same label.

Note that each caption ends with a well-known example of the word under discussion. These examples are set in italics. Also set in italics are other geographical terms that are related, in some way, to the word which is being defined.

ARCHIPELAGO A
ATOLL B
BAY C
CANYON D
CAPE E
CONTINENTAL DIVIDE F
DELTA G
ESTUARY H
FJORD I
HEADWATERS J
GLACIER K
GULF L
ISLAND M
ISTHMUS N
LAGOON O
MESA P
OCEAN CURRENTS Q
PENINSULA R
PLATEAU S
REEF T
STRAIT U

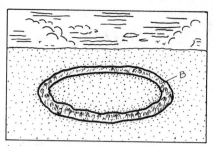

A circular coral island that encloses a *lagoon*. Atolls are usually formed on top of submerged *volcanoes*. *Bikini Atoll in the Marshall Islands of the Pacific Ocean, a US atomic test site.*

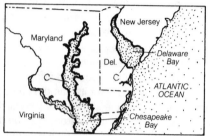

A body of water that penetrates a coastline. It is generally wider in the middle. It is usually smaller than a *gulf*, but larger than a *cove*. *Delaware and Chesapeake Bays.*

A deep, narrow depression in the earth's surface, often having a river running through it. Canyons are also known as *gorges*. *Ravines* are not quite as deep. *The Grand Canyon in northwest Arizona.*

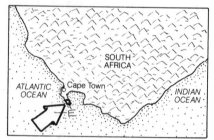

A point of land extending into the sea. It is usually smaller than a *peninsula*. A mountainous cape is called a *promontory* or a *headland*. *The Cape of Good Hope off the South African coast.*

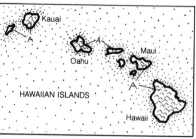

Either a group of islands or a body of water that has many islands in it. *The Hawaiian Islands; the Aegean Sea off the coast of Greece.*

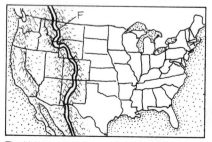

The highest point of a continent, from which the direction of river flow is determined. *The Great Divide is the name given to the crest of the Rocky Mountains, which sends rivers east and west.*

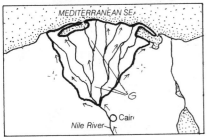

The triangular-shaped land found at the mouth of some large rivers. So much soil is transported by the river that the coastal waters cannot wash it all away. *The Nile Delta on the Mediterranean Sea.*

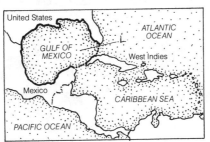

A part of an ocean or sea that is partially enclosed by a curving coastline. A more fully enclosed body of salt water could be called a *sea*. *The Gulf of Mexico.*

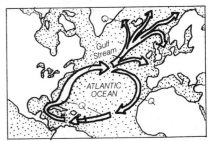

Ocean streams that are propelled by prevailing winds and earth rotation. They flow clockwise in the northern hemisphere and counterclockwise below the equator. *The Atlantic's Gulf Stream.*

An ocean inlet that merges with the mouth of a river. The estuary's salinity varies according to river flow and ocean tides. *The Río de la Plata, separating Argentina from Uruguay.*

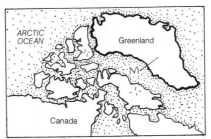

A body of land completely surrounded by water. It is smaller than a *continent* but larger than a *cay*, a *key*, or certainly a *large rock*. *Greenland is the world's largest island.*

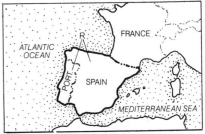

A mass of land almost entirely surrounded by water. It is usually connected to the mainland by a narrow neck. *The Iberian Peninsula in Europe, home to Spain and Portugal.*

A narrow, winding ocean inlet that penetrates a coastal mountain range. The steep cliffs that line its route make a fjord (fiord) one of nature's grandest sights. *Norway's Sogne Fjord is the world's longest.*

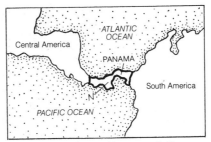

A narrow strip of land, with water on both sides, that connects two larger land masses. *The Isthmus of Panama connects Central America and South America.*

A broad expanse of generally high and flat land, also called a *tableland*. Plateaus can rise up from a lower area, or can be level regions within a mountain range. *Most of Spain is the Meseta Plateau.*

Upper river springs, streams, and tributaries. Headwaters can refer to *continental divides* or *watersheds*. *Watershed* also describes a region drained by a river. *The Alps have been called the headwaters of Europe.*

A small body of water separated from the larger sea by a barrier of *sand or coral reefs*. It can either be adjacent to a coastline or surrounded by an *atoll*. *Mirim Lagoon off the coast of Brazil.*

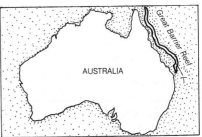

A narrow, low ridge of rock, or more commonly of coral, that is connected to a coast *(fringing reef)* or lies off a coast *(barrier reef)*. The Great Barrier Reef, off the northeast coast of Australia.

A river of ice, moving slowly down a mountain slope or outward from its central mass. It stops where the leading edge melts faster than the forward rate of movement. *Vatnajökull in Iceland is Europe's largest.*

A tall, flat-topped mountain with steep vertical sides. Erosion-resistant mesas are left standing after all else has gone. *Buttes* are small mesas. *Monument Valley in Utah has 1,000 ft. (305 m) mesas.*

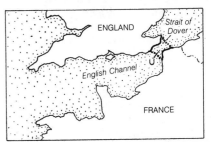

A narrow passage of water connecting two larger bodies of water. A *channel* is wider than a *strait*. If it is shallow, it is called a *sound*. *The English Channel becomes narrower at the Strait of Dover.*

# CONTINENTS OF THE WESTERN HEMISPHERE *

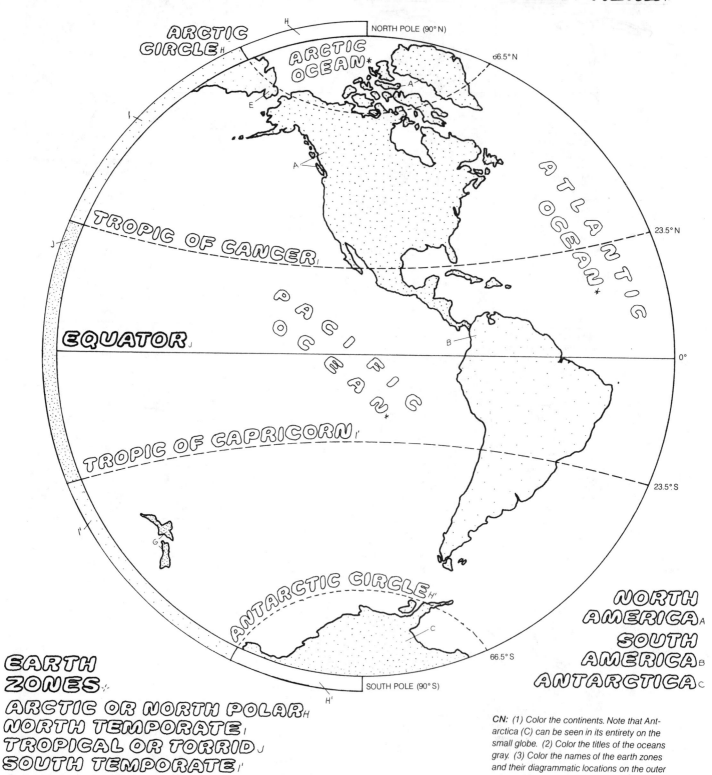

**EARTH ZONES**
**ARCTIC OR NORTH POLAR** H
**NORTH TEMPORATE** I
**TROPICAL OR TORRID** J
**SOUTH TEMPORATE** I'
**ANTARCTIC OR SOUTH POLAR** H'

**NORTH AMERICA** A
**SOUTH AMERICA** B
**ANTARCTICA** C

*CN: (1) Color the continents. Note that Antarctica (C) can be seen in its entirety on the small globe. (2) Color the titles of the oceans gray. (3) Color the names of the earth zones and their diagrammatic locations on the outer edges of the maps.*

Earth zones are defined by imaginary lines of latitude circling the globe, parallel to the equator. The latitude lines shown above are not parallel because of the type of map projection; they are parallel in the global view to the right. The *tropical or torrid zone* is the largest and hottest. The sun is always shining directly over some part of it. It is bounded by the *tropics of Cancer* and *Capricorn*. The *equator* passes through the center of the tropical zone, halfway between the two poles, dividing the earth into northern and southern hemispheres. The northern boundary of the tropical zone is the tropic of Cancer, the northernmost parallel (23.5°N latitude) where the sun shines directly overhead (noon of the summer solstice). About 75% of the earth's population lives in the *north temperate zone*. The four seasons occur only in the temperate zones. The *arctic circle*, 23.5° from the pole, is the southern boundary of the *arctic* or *north polar zone*. In this zone the sun fails to rise during the winter months. The sun stays below the horizon for one day at the arctic circle—and for six months at the north pole. During the summer, the sun fails to set for a comparable period of time (see Plate 41). The tropic of Capricorn is the southern boundary of the tropical zone and the southernmost parallel (23.5°S) where the sun appears overhead (noon of the winter solstice). This line also marks the northern border of the *south temperate zone,* which is limited to the south by the *antarctic circle* (23.5° from the south pole); this in turn is the northern boundary of the *antarctic zone or south polar zone or region.*

# CONTINENTS OF THE EASTERN HEMISPHERE*

EUROPE D
ASIA E
AFRICA F
OCEANIA G

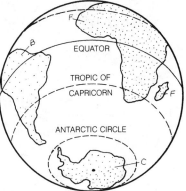

*Continents* are large land masses with adjacent islands, surrounded, or nearly surrounded by water. The seven continents cover slightly less than 30% of the earth's surface. The rest of our "water planet" is covered by four oceans and many *seas* (shallower extensions of oceans, partially surrounded by land).

| CONTINENT | % OF TOTAL LAND | LAND AREA | % OF TOTAL POPULATION | POPULATION |
|---|---|---|---|---|
| Asia | 29.5 | 17,230,000 sq.mi. (44,625,700 km²) | 63.0 | 3,200,000,000 |
| Africa | 20.0 | 11,700,000 sq.mi. (30,279,600 km²) | 10.6 | 600,000,000 |
| North America | 16.3 | 9,400,000 sq.mi. (24,346,680 km²) | 7.7 | 450,000,000 |
| South America | 11.8 | 6,900,000 sq.mi. (17,871,400 km²) | 5.5 | 300,000,000 |
| Antarctica | 9.6 | 5,400,000 sq.mi. (13,986,000 km²) | — | — |
| Europe | 6.5 | 3,810,000 sq.mi. (9,867,900 km²) | 12.9 | 700,000,000 |
| Oceania | 5.2 | 3,300,000 sq.mi. (8,547,000 km²) | 0.3 | 16,800,000 |

| OCEAN | % OF TOTAL OCEAN | OCEAN AREA | MAXIMUM DEPTH |
|---|---|---|---|
| Pacific | 49.2 | 64,100,000 sq. mi. (165,890,800 km²) | 36,170 ft. (11,027 m) |
| Atlantic | 24.6 | 32,220,000 sq. mi. (83,385,360 km²) | 30,200 ft. (9,207 m) |
| Indian | 22.0 | 28,900,000 sq. mi. (74,793,200 km²) | 24,440 ft. (7,451 m) |
| Arctic | 4.2 | 5,300,000 sq. mi. (13,716,400 km²) | — |

# MOVEMENT OF CONTINENTS

NORTH AMERICA A
SOUTH AMERICA B
ANTARCTICA C
EUROPE D
ASIA E
AFRICA F
OCEANIA G

*CN: Use the same colors you used for the continents on Plate 1. Use a very light color for the earthquake zones (I). (1) Color the four maps on this page; complete each one before doing the next. (2) At the top of the opposite page, color the tiny triangles representing volcano sites. (3) Color the earthquake zones (areas covered with light parallel lines).*

## 200 MILLION YEARS AGO

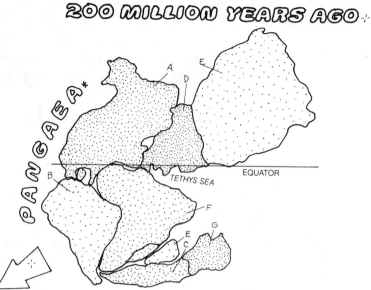

Around 200 million years ago there was a single continent called Pangaea (Greek for "all earth"). The Tethys Sea, ancestor of the Mediterranean, partially divided this land mass. It is assumed that the continents making up Pangaea previously migrated from other unknown locations.

## 100 MILLION YEARS AGO

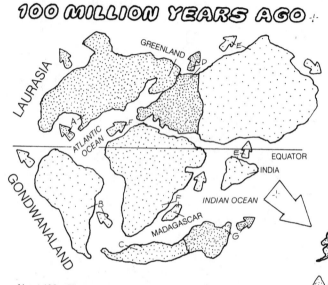

About 100 million years ago, Pangaea divided into two masses: Laurasia (North America, Europe, and Asia) and Gondwanaland (South America, Africa, Australia, and Antarctica). The Atlantic and Indian Oceans were expanding. India was heading for the Asian continent. The continents were all moving northward, and Greenland was starting to break away from North America.

## TODAY

Today we find Australia far from Antarctica. India is now an integral part of Asia. The Atlantic Ocean is still expanding, and Africa and South America are well within the equatorial region. Note how far north of the equator the continents have drifted. This explains why evidence of past tropical vegetation is found in northern regions.

## 50 MILLION YEARS FROM TODAY

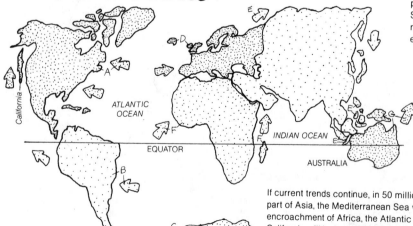

If current trends continue, in 50 million years Australia will be part of Asia, the Mediterranean Sea will be greatly reduced by the encroachment of Africa, the Atlantic will be wider, and part of California will be headed for Alaska.

# VOLCANO SITES

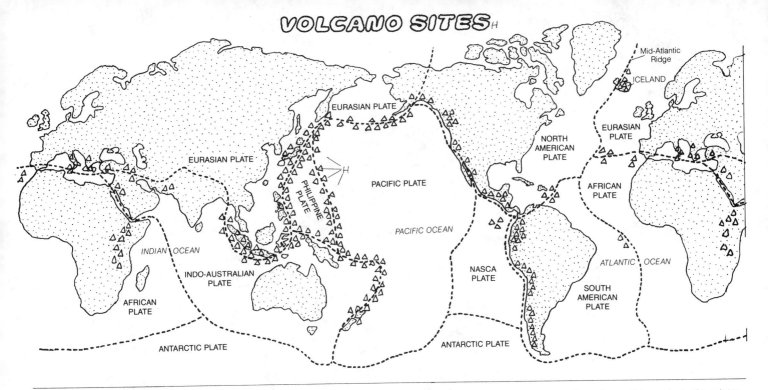

For many years scientists wondered why coal deposits, the decay products of tropical vegetation, were located in northern regions. Why was there evidence of glacial formations in tropical Africa? Why did the opposing coastlines of South America and Africa look as if they could fit together?

Early in this century, Alfred Wegener, a German scientist, attempted to explain these phenomena. According to his theory, the continents have changed locations over millions of years and they are still in motion because they rest on moving sections of the earth's crust called *tectonic plates*. These plates act as huge rafts floating on the earth's mantle, which in turn is moved by convection currents deep within it. The crust is 19–25 miles (30–40 km) thick; the mantle is about 100 times thicker. The 9 plates shown on this page are the largest of 18 known plates. Plate movement generally follows major earthquakes; it averages 3/4 in. (2 cm) to 2 in. (5 cm) per year. Cracks in the mantle allow magma (molten rock) to rise to the surface in volcanic eruptions, and new plate material is created. A major site for this process is the Mid-Atlantic Ridge near Iceland. As the newly formed North American and Eurasian plates move away from the ridge, they expand the Atlantic Ocean.

Moving plates have to go somewhere; they often collide against other plates. When such collisions occur, a crumpled plate will be the source of new mountain material. The tallest mountains in the world, the Himalayas in northern India, are still growing as the Indo-Australian plate pushes under the Eurasian plate. Plates can also slide past each other; the junction between the moving plates is called a *fault*. The San Andreas fault, which knifes along the coast of California, separates the Pacific plate (which carries San Francisco and Los Angeles) from the North American plate. In 30 million years, the Pacific plate will move 400 miles (640 km) northward and Los Angeles will replace San Francisco opposite Oakland in the fault-divided San Francisco Bay Area. In 50 million years, Los Angeles will be on its way to Alaska (see lower map on opposite page).

Events along the borders of moving plates release enormous energy in the form of earthquakes and volcanoes. The coastlines and islands on the rim of the Pacific Ocean are in a violently active geological region called the "Ring of Fire." Another major earthquake and volcano zone runs westward from Southeast Asia through China, the Middle East, and southern Europe.

# EARTHQUAKE ZONES

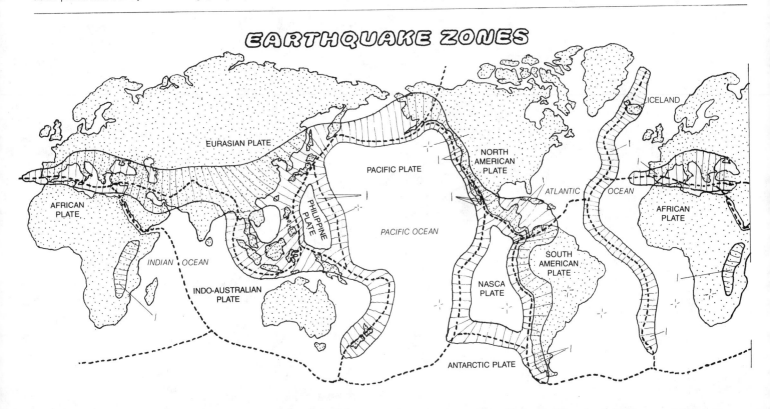

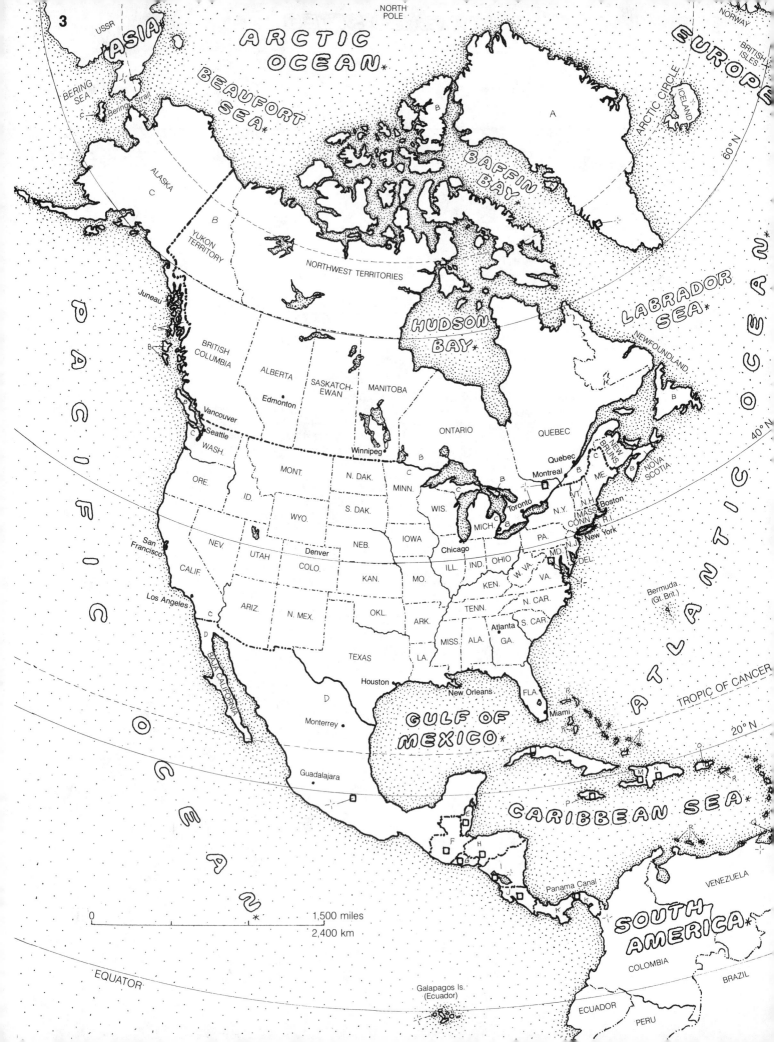

# NORTH AMERICA: THE COUNTRIES

CN: (1) Use very light colors for Canada (B) and the US (C). Do not color the squares representing the capital cities. (2) Note that only the larger West Indian island-nations are named and colored separately (L–Q). The smaller ones are all labeled "Other Islands" (R) and are colored alike (see Plate 14 for the identification of those islands). (3) Color North America gray on the global map. Color the colonial map below.

GREENLAND A / GODTHÅB (DENMARK)
CANADA B / OTTAWA
UNITED STATES C / WASHINGTON, D.C.
MEXICO D / MEXICO CITY

## CENTRAL AMERICA
BELIZE E / BELMOPAN
GUATEMALA F / GUATEMALA CITY
EL SALVADOR G / SAN SALVADOR
HONDURAS H / TEGUCIGALPA
NICARAGUA I / MANAGUA
COSTA RICA J / SAN JOSÉ
PANAMA K / PANAMA CITY

## WEST INDIES
CUBA L / HAVANA
HAITI M / PORT-AU-PRINCE
DOMINICAN REPUBLIC N / SANTO DOMINGO
PUERTO RICO O / SAN JUAN (US)
JAMAICA P / KINGSTON
TRINIDAD & TOBAGO Q / PORT OF SPAIN
OTHER ISLANDS R

North America stretches from deep within the arctic circle to close to the equator. Note that the continent lies almost entirely west of South America.

The map below shows the state of European domination just before France lost its mainland possessions. France gave up its lands east of the Mississippi to England and sold its midwestern holdings in the Louisiana Purchase of 1803, which doubled the size of the US. The US went on to gain additional land from Spain, Mexico, and Russia by means of war, treaty, and purchase.

North America, which includes Greenland, Central America, and the West Indies, covers 9,400,000 square miles (24,346,000 km²). It is the third largest continent (after Asia and Africa). Its population of 450 million is the fourth largest (after Asia, Europe, and Africa). Canada, the continent's largest country, has only a tenth the population of the United States. Population density in North America increases the further south one goes.

Most North Americans are descended from people who immigrated within the past 400 years from all other continents except Antarctica. The majority came from Europe. The true natives, named "Indians" by Columbus, who thought he had landed in the East Indies, numbered about 5 million. Their ancestors came from Asia 20,000–40,000 years ago, crossing a land bridge uncovered during the last Ice Age, when the formation of ice lowered the sea. The bridge is now under the Bering Strait. Eskimos made the same journey around 6,000 years ago. Today, most people living south of the US are of mixed European and Indian ancestry, and are called "Mestizos." Blacks living in the US and the West Indies are descended from slaves brought from Africa.

The Vikings were the first Europeans to set foot on North America. After Columbus's arrival some 500 years later, exploration, colonization, and exploitation began in earnest. In the early 1500s the Spanish, with superior weaponry and military cunning, easily conquered the natives of the southwestern United States, Mexico, Central America, and the West Indies. A century later, the French explored Canada and the central US as far south as New Orleans. Both nations introduced the Roman Catholic religion. English Protestants colonized northern Canada and the eastern seaboard of the US, the site of the 13 original colonies. Though their rule ended some 200 years ago, the English left a permanent stamp on the American language, religion, and culture. Along with the French and the Dutch, the British continue to control many islands in the West Indies, the last vestiges of colonization of North America.

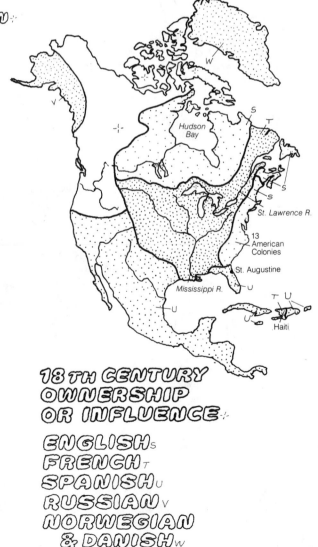

## 18TH CENTURY OWNERSHIP OR INFLUENCE

ENGLISH S
FRENCH T
SPANISH U
RUSSIAN V
NORWEGIAN & DANISH W

**CN:** You will probably need to use some colors more than once.
Use your lightest colors for S–X and 2–6 (the land regions to the right).
(1) Use one color for all five Great Lakes (O); color Great Slave Lake
(G¹) and Lake Winnipeg (P¹) the same color as the rivers that fill them.
(2) Use gray for the mountain peaks on the right and the names of
the bodies of water on the large map.

## CANADIAN SHIELD 2.
## EASTERN UPLANDS 3.
## CENTRAL PLAINS 4.
## WESTERN MOUNTAINS 5.
## COASTAL LOWLANDS 6.

## PRINCIPAL RIVERS -:-
ARKANSAS A
CHURCHILL B
COLORADO C
COLUMBIA D
FRAZER E
MACKENZIE F
SLAVE G
PEACE H
MISSISSIPPI I
MISSOURI J
NELSON K
OHIO L
RED M
RIO GRANDE N
ST. LAWRENCE O
N. & S. SASKATCHEWAN P
SNAKE Q
YUKON R

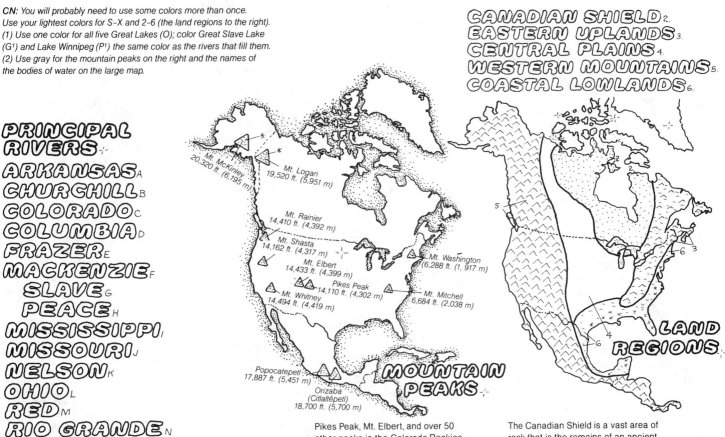

Pikes Peak, Mt. Elbert, and over 50 other peaks in the Colorado Rockies are higher than 14,000 ft. (4,268 m). Mt. McKinley is the tallest peak on the continent. Popocatepetl is the tallest of the many active volcanoes in Mexico and Central America.

The Canadian Shield is a vast area of rock that is the remains of an ancient, completely worn down range. It is a storehouse of largely untapped minerals. The thin layer of soil covering it can support forests but not farming. Hudson Bay fills a large depression in the Shield.

## PRINCIPAL MOUNTAIN RANGES -:-
APPALACHIAN MTS. S
CASCADE RANGE T
COAST MTS. U
ROCKY MTS. V
SIERRA MADRE (E) W (W) W'
SIERRA NEVADA X

## PRINCIPAL LAKES -:-
### THE GREAT LAKES -:-
L. SUPERIOR O¹
L. HURON O²
L. MICHIGAN O³
L. ERIE O⁴
L. ONTARIO O⁵
GREAT BEAR L. Y
GREAT SLAVE L. G¹
GREAT SALT L. Z
L. WINNIPEG P¹
L. NICARAGUA 1.

North America, the continent with the longest coastline, has a widely varied landscape and climate: frozen wastes in Greenland and northern Canada, evergreen forests in central Canada and the western US, towering peaks in the Rocky Mountains, barren deserts in the southwestern US, palm-covered islands in the Caribbean Sea, steamy jungles in Central America. Even the shape of the continent is unusual: a broad expanse of 4,900 miles (7,840 km) from Alaska to Newfoundland narrows to a mere 30 miles (48 km) at the Isthmus of Panama. Variations in temperature between the polar north and the tropical south can exceed 200° F (93° C). Parts of Greenland (85% of which is covered with ice) and the Yukon have recorded temperatures as low as -105° F (-76° C). Death Valley, California, is the lowest point on the continent at 282 feet (86 m) below sea level; it has baked in 134° F (57° C) heat.

The continent has three basic physical regions: an eastern uplands of very old, low, worn-down mountains; a much younger, steeper, and more rugged series of western mountain ranges that cover a third of the continent, from Alaska to Central America; and a broad, flat area in between, which is further broken down into the central plains, Canadian Shield, and coastal lowlands areas (see map, upper right).

There are three large river systems. (1) The Mississippi-Missouri complex forms the continent's longest river at 3,872 miles (6,195 km), flowing south to the Gulf of Mexico. (2) The Great Lakes (including Lake Superior, the world's largest freshwater lake) and the St. Lawrence River and Seaway flow to the Atlantic Ocean. (3) The Mackenzie and Nelson systems flow to northern Canadian waters. Most river flow in North America is directed by the crest of the Rocky Mountains, the Great Divide. (Such a watershed region is called a continental divide.) North American rivers flow west to the Pacific, east to the Atlantic, north to Hudson Bay or the Arctic Ocean, or south to the Gulf of Mexico, depending on which side of the crest they originate from.

# NORTH AMERICA: CANADA & GREENLAND

**Area:** 3,850,000 sq. mi. (9,971,500 km²). **Population:** 26,500,000. **Capital:** Ottawa, 720,000. **Government:** Constitutional monarchy. **Language:** English 65%; French 20%; 15% speak both. **Religion:** Roman Catholic 47%; Protestant 40%. **Exports:** Timber, newsprint, autos, machinery, fish products, grains, asbestos, nickel, and zinc. **Climate:** West coast is mild; southern and eastern regions have warm summers and very cold winters; north is frigid. □ Canada, which is about the size of Europe, is the world's second largest country (after the USSR). It spans North America from the Atlantic to Alaska and shares with the United States the world's longest unprotected border. Though 7% larger than the US, Canada has only a tenth the population. Because of the cold climate, most Canadians live within a 200-mile (320 km) strip along the southern border with the US.

Even with the growth of minority groups, Canada remains a nation of two cultures: English and French (both are official languages). Though France gave up its Canadian holdings to England in 1763, the French minority has always resisted assimilation, and today, separatist groups are moving French-dominated Quebec toward complete independence.

Canada has vast undeveloped resources. A wide band of forests spanning the nation enables Canada to be the world's leading producer of pulp and paper. It has the most lakes and rivers of any country (about a third of the world's freshwater supply), which provide transportation routes, irrigation, and hydroelectric power. Canada shares four of the five Great Lakes with the US. It has 9 lakes over 100 miles (160 km) long, 35 that are more than 50 miles (80 km) long. The Mackenzie-Peace is the longer of the two major river systems, but far more important is the St. Lawrence River Seaway network, which gives Atlantic Ocean access to Canada's population and industrial centers, from the Great Lakes to the eastern seaboard.

Canada is an independent federation of 10 self-governing provinces and 2 territories, but Canadians regard the queen of England as their queen. Canada belongs to the English Commonwealth of Nations, a worldwide association of former colonies and current dependencies. Canada consists of six regions. (1) The Atlantic or Maritime Provinces are *New Brunswick, Newfoundland, Nova Scotia,* and *Prince Edward Island.* The smallest provinces were settled first. The Bay of Fundy, separating New Brunswick and Nova Scotia, is famous for its 70 foot (21 m) tides. The scenic fjords on the coast of Newfoundland must have looked like home to the Viking explorers, who in 1000 AD established a short-lived colony called Vinland. (2) *Quebec* is the largest province and the only one in which the French language and the Roman Catholic religion represent the majority. More French speaking people live in Montreal than in any other city except Paris. Quebec is almost entirely covered by the Canadian Shield (see diagram on Plate 4). Mining and timber are the major industries. (3) *Ontario,* the site of Ottawa, the Canadian capital, is the most populous and industrialized province. Ontario's capital, Toronto, has 3 million people and is Canada's center of commerce and manufacturing. The many mines of Ontario contribute to Canada's world leadership in nickel and zinc production. (4) The Prairie Provinces—*Alberta, Saskatchewan,* and *Manitoba*—are home to enormous wheat farms and cattle ranches. Alberta has the world's largest national park (Wood Buffalo) and two of the most scenic (Banff and Jasper). Winnipeg, the "Chicago" of Canada, is the capital of Manitoba and the major transportation hub. (5) *British Columbia,* Canada's Pacific coast, is the most beautiful province, with its thick forests, snowcapped mountains, and fjord-lined coast. Timber and fishing are the major industries. Vancouver is the nation's busiest seaport. (6) The *Yukon* and *Northwest Territories* occupy a third of the land area but have only 1% of the people. About 25,000 Eskimos (called Inuit in Canada) live in small communities in the arctic region. About 370,000 native Indians live to the south, across the entire nation. The north's brutal climate has not prevented extensive mining and oil exploration. Gold made the Klondike region of the Yukon famous but is no longer the region's most valued mineral.

Prosperous Canada has vast natural resources and unlimited land for population expansion. The standard of living is equal to that of the US. But just as the French-speaking minority strives to preserve its identity, Canadians as a whole are fearful of being swallowed up culturally and economically by their neighbor to the south.

---

**CN:** (1) Color a province or territory, its name, and its outline on the large map. (2) Color the outline of Greenland on the large map and then on the globe to the right (on which Canada and its northern islands are colored gray).

## PROVINCES

ALBERTA <sub>A</sub> EDMONTON

BRITISH COLUMBIA <sub>B</sub> VICTORIA

MANITOBA <sub>C</sub> WINNIPEG

NEW BRUNSWICK <sub>D</sub> FREDERICTON

NEWFOUNDLAND <sub>E</sub> ST. JOHN'S

NOVA SCOTIA <sub>F</sub> HALIFAX

PRINCE EDWARD I <sub>G</sub> CHARLOTTETOWN

ONTARIO <sub>H</sub> OTTAWA

QUEBEC <sub>I</sub> QUEBEC

SASKATCHEWAN <sub>J</sub> REGINA

## TERRITORIES

NORTHWEST TERRITORIES <sub>K</sub> YELLOWKNIFE

YUKON TERRITORY <sub>L</sub> WHITEHORSE

## GREENLAND <sub>M</sub>

**Area:** 840,000 sq. mi. (2,175,600 km²). **Population:** 55,000. **Capital:** Godthåb, 10,750. **Government:** Self-governing province of Denmark. **Language:** Danish; Eskimo dialect. **Exports:** Fish products. **Climate:** Frigid, but summers are above freezing. □ Greenland is the world's largest island and the coldest nation. The population is confined to the southwest coast, which is not as cold as the ice-covered interior. Mountains fringe a giant sheet of ice, 1–2 miles (1.6–3.2 km) thick, which covers 85% of the island. The glacial ice, created under the weight of many layers of accumulated snow, is pushed into the sea, where it breaks up into the icebergs that menace North Atlantic shipping. Greenlanders are descendants of Canadian Eskimos and Danish settlers (Vikings) who arrived in 982. Eric the Red coined the name "Greenland" to entice immigrants from home. He may have been misled by the coastline, which turns green during the cool summer. The generally frigid weather is at least partially responsible for the world's highest suicide rate. When the fish aren't running, unemployment is rampant. There is hope that the recent discovery of gold will bring prosperity.

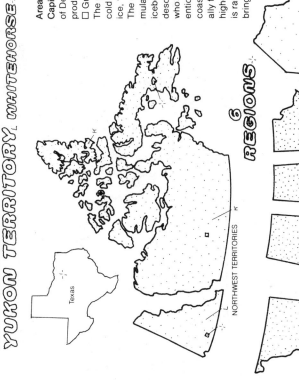

REGIONS

NORTHWEST TERRITORIES

Texas

ONTARIO

QUEBEC

ATLANTIC PROVINCES

BRITISH COLUMBIA

PRAIRIE PROVINCES

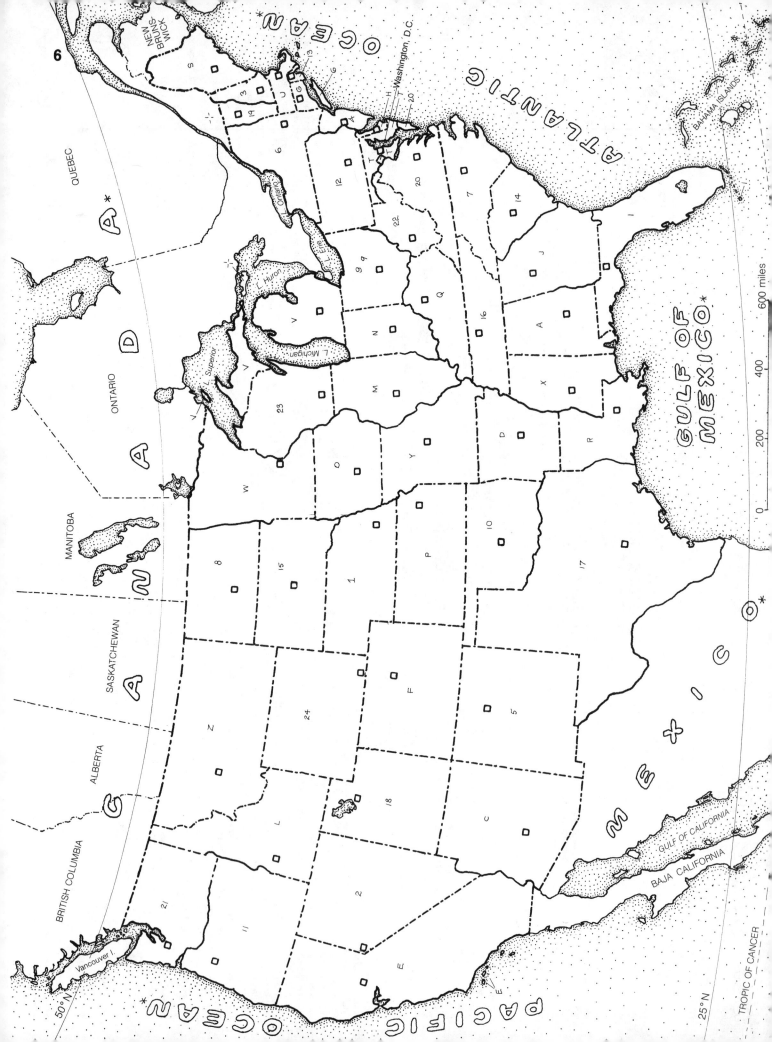

# NORTH AMERICA: UNITED STATES

**Area:** 3,679,204 sq. mi. 9,529,138 km². **Population:** 247,000,000. **Capital:** Washington, D.C. 638,500. **Government:** Republic. **Language:** English; many ethnic dialects. **Religion:** Protestant 33%; Roman Catholic 22%. **Exports:** Machinery, aircraft, electronics, coal, iron and steel, chemicals, textiles, cotton, grains, soybeans, corn, and produce. **Climate:** Temperate, except for northern and midwestern extremes. Pacific coast is mild. Southwest is dry with very hot summers. Gulf coast is subtropical □ During the 20th century the United States became the world's most affluent and powerful nation. It is fourth in size (after the USSR, Canada, and China), and fourth in population (after China, India, and the

USSR). The US has benefited from enormous natural resources; an ideal location for trade; vast regions of fertile land and favorable weather; friendly neighbors and protective ocean barriers; a classless social and economic system that rewards individual effort; a stable and democratic government; and from the existence of personal, religious, and economic freedom for its people, many of whom are highly motivated immigrants. At the peak of US prosperity, the value of its manufactured, mined, and grown products far exceeded that of any other nation—6% of the world's population was producing 50% of its wealth. In the 1970s, the US began to face growing competition from Europe and Asia. In addition, the cost of waging war, both hot and cold, against international communism had drained its resources. Worst of all, years of uninterrupted prosperity had begun to weaken the sense of purpose that has sustained American society. The 1990s find the US, once the world's largest creditor, now the largest debtor nation, and its preeminent economic position is eroding at an ever-increasing rate.

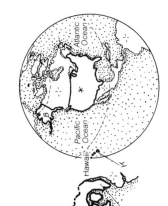

ARCTIC OCEAN

CANADA

BERING SEA

Bering Strait

USSR

Pacific Ocean

Atlantic Ocean

Hawaii

ALABAMA A / MONTGOMERY
ALASKA B / JUNEAU
ARIZONA C / PHOENIX
ARKANSAS D / LITTLE ROCK
CALIFORNIA E / SACRAMENTO
COLORADO F / DENVER
CONNECTICUT G / HARTFORD
DELAWARE H / DOVER
FLORIDA I / TALLAHASSEE
GEORGIA J / ATLANTA
HAWAII K / HONOLULU
IDAHO L / BOISE
ILLINOIS M / SPRINGFIELD
INDIANA N / INDIANAPOLIS
IOWA O / DES MOINES
KANSAS P / TOPEKA
KENTUCKY Q / FRANKFORT
LOUISIANA R / BATON ROUGE
MAINE S / AUGUSTA
MARYLAND T / ANNAPOLIS
MASSACHUSETTS U / BOSTON
MICHIGAN V / LANSING
MINNESOTA W / ST. PAUL
MISSISSIPPI X / JACKSON
MISSOURI Y / JEFFERSON CITY
MONTANA Z / HELENA
NEBRASKA 1 / LINCOLN
NEVADA 2 / CARSON CITY
NEW HAMPSHIRE 3 / CONCORD
NEW JERSEY 4 / TRENTON
NEW MEXICO 5 / SANTE FE
NEW YORK 6 / ALBANY
NORTH CAROLINA 7 / RALEIGH
NORTH DAKOTA 8 / BISMARCK

OHIO 9 / COLUMBUS
OKLAHOMA 10 / OKLAHOMA CITY
OREGON 11 / SALEM
PENNSYLVANIA 12 / HARRISBURG
RHODE ISLAND 13 / PROVIDENCE
SOUTH CAROLINA 14 / COLUMBIA
SOUTH DAKOTA 15 / PIERRE
TENNESSEE 16 / NASHVILLE
TEXAS 17 / AUSTIN
UTAH 18 / SALT LAKE CITY
VERMONT 19 / MONTPELIER
VIRGINIA 20 / RICHMOND
WASHINGTON 21 / OLYMPIA
WEST VIRGINIA 22 / CHARLESTON
WISCONSIN 23 / MADISON
WYOMING 24 / CHEYENNE

# NORTH AMERICA: NORTHEASTERN U.S.

All the northeastern states were among the original 13 British colonies. The New England states are Connecticut, Maine, Massachusetts, New Hampshire, Rhode Island, and Vermont. Their historic past is still evident in the names, monuments, schools, museums, and architecture of the area. The region's famous autumn foliage is unsurpassed in brilliance of color. The northernmost states have particularly long and cold winters.

Adjacent to New England are the mid-Atlantic states, New York, New Jersey, and Pennsylvania. This major industrial region has the greatest population concentration in the nation. A megalopolis of major cities runs from Boston to Washington, D.C. Further south are the culturally Southern former slave states of Delaware and Maryland. They remained in the Union during the Civil War, but loyalty among their populations was sharply divided.

Mountains that cover much of the northeast are part of the ancient Appalachian chain, whose peaks have been worn smooth. The only flatland is the Atlantic coastal plain of southeastern New Jersey, eastern Maryland, and Delaware.

## CONNECTICUT<sub>A</sub>

**Area:** 5,009 sq. mi. (12,963 km²). **Population:** 3,230,000. **Capital:** Hartford, 137,000. **Economy:** Aircraft engines, nuclear submarines, tobacco, dairying, and insurance. ☐ The Connecticut (don't pronounce the middle "c") River is New England's longest. It separates Vermont from New Hampshire and flows south, through Massachusetts and Connecticut, on its way to Long Island Sound. In wooded and hilly Connecticut, the river valley is a low, fertile region where the most expensive tobacco (for wrapping cigars) is shade-grown. Connecticut was known as the birthplace of many revolutions and innovations that contributed to both the industrial and American revolutions; an industrial area around Waterbury was called the "arsenal of the nation." Hartford, second in size to Bridgeport (143,000), is the nation's insurance industry capital. Yale University is located in New Haven. Many residents in the affluent southwest corner of the state commute to New York City.

## DELAWARE<sub>B</sub>

**Area:** 2,045 sq. mi. (5,292 km²). **Population:** 660,000. **Capital:** Dover, 23,600. **Economy:** Chemicals, poultry, food products, and fishing. ☐ Sharing the Delmarva Peninsula with Maryland and Virginia, this second smallest state spans less than 35 mi. (56 km) at its widest point. Delaware is the state with the lowest elevation. The only hills are in the northern industrial region of Wilmington (70,200), the largest city. Home of Du Pont, the chemical giant which has dominated Delaware economically and politically for nearly two centuries, Wilmington has been called the "chemical capital of the world." Delaware is known for liberal laws regulating business. Many US companies (half of the top 500), though located elsewhere, are Delaware corporations. In 1638, Delaware was the Swedish colony of "New Sweden"; it was taken over by the Dutch and then the British, who made it part of Pennsylvania.

## MAINE<sub>C</sub>

**Area:** 33,260 sq. mi. (86,076 km²). **Population:** 1,220,000. **Capital:** Augusta, 21,850. **Economy:** Paper products, fishing, potatoes, blueberries, and ship-building. ☐ The northern half of Maine, the most easterly state, is completely surrounded by Canada. Maine is a major producer of paper products; 90% of the state is covered by forests, most owned by paper companies. Hordes of visitors come to the scenic rocky coast each summer. Mt. Desert Island is the site of Acadia, New England's only national park. Maine has so many islands that its name refers to the "mainland." The busy port of Portland (62,000) is the largest city. Potatoes are the principal crop, and the famed "Maine lobster" is the prize catch of the fishing industry.

## MARYLAND<sub>D</sub>

**Area:** 10,470 sq. mi. (27,096 km²). **Population:** 4,735,000. **Capital:** Annapolis, 32,000. **Economy:** Trade, electronics, chemicals, food products, tobacco, and poultry. ☐ The Chesapeake Bay, a bountiful provider of seafood (particularly crabs, clams, and oysters), nearly divides Maryland in two. The eastern shore is a flat, fertile vegetable-producing area. The larger western shore is densely populated, especially near Baltimore and Washington D.C. The land becomes progressively hillier toward the Appalachians. Baltimore (785,000) is one of the nation's busiest seaports and is a center of trade and manufacturing. The city is famous for endless rows of attached brick houses with white marble steps. Nearby Annapolis, the state capital, is the site of the US Naval Academy. The northern boundary of Maryland was the Mason-Dixon line, which divided the slave states of the South from the abolitionist North.

## MASSACHUSETTS<sub>E</sub>

**Area:** 8,257 sq. mi. (21,369 km²). **Population:** 5,930,000. **Capital:** Boston, 553,000. **Economy:** Electronics, precision products, fishing, textiles, and cranberries. ☐ Small only in size, Massachusetts is a leader in finance, trade, industry, culture, medicine, and education—almost all of which are centered in the greater Boston area, along with half the population. Here is the site of Harvard, the nation's oldest university, and many other leading colleges, including the Massachusetts Institute of Technology. Boston was founded in 1630, only 10 years after the Pilgrims landed at Plymouth Rock on Cape Cod Bay. Colonial architecture, museums, monuments, and restored villages are visible reminders of the state's historic past. The Cape Cod peninsula and the islands of Martha's Vineyard and Nantucket are popular summer resorts on the Atlantic coast. Massachusetts is no longer a major producer of clothing, textiles, and shoes. The mills and factories have largely been replaced by electronics and engineering companies.

## NEW HAMPSHIRE<sub>F</sub>

**Area:** 9,380 sq. mi. (24,275 km²). **Population:** 1,105,000. **Capital:** Concord, 30,500. **Economy:** Wood products, electronics, granite, shoes, and dairying. ☐ The "Granite State" is known for its large deposits of many varieties of granite. New Hampshire is almost completely mountainous except in the southeastern Merrimack River valley, in which Concord, the capital, and Manchester (91,000), the largest city, are located. The state barely has a coastline—only 13 mi. (21 km) separate Maine from Massachusetts. In the dramatic White Mountains of the north is the tallest peak in the northeast, Mt. Washington (6,288 ft., 1,917 m). Around its summit swirled the highest winds ever clocked: 231 mph (370 km/h).

## NEW JERSEY<sub>G</sub>

**Area:** 7,790 sq. mi. (20,160 km²). **Population:** 7,620,000. **Capital:** Trenton, 92,200. **Economy:** Chemicals, pharmaceuticals, food products, and fishing. ☐ The "Garden State" is named for its flower-filled hot houses, fruit orchards, and productive truck farms. But New Jersey is also a major industrial state. It has the nation's highest population density and the greatest percentage (95%) of town and city dwellers. Many residents commute to Philadelphia and New York. The state's largest city is Newark (385,000). New Jersey is the shipping hub of the Northeast and its ports on New York Bay handle almost all the harbor's container traffic. Atlantic City, with its gambling casinos, heads the list of the state's popular seaside resorts.

## NEW YORK<sub>H</sub>

**Area:** 52,730 sq. mi. (136,465 km²). **Population:** 17,600,000. **Capital:** Albany, 101,780. **Economy:** Manufacturing, finance, trade, clothing, produce, and dairying. Though overtaken by California in population and manufacturing output, New York remains the national leader in business and trade. New York City, the "Big Apple," with over 7 million people, is the business capital of the US, if not the world. It is also the center of finance, trade, fashion, advertising, publishing, music, theater, and the arts. New York's dominance in commerce began with the opening of the Erie Canal in 1825. The canal, linked to the Mohawk and Hudson Rivers, gave midwestern cities on the Great Lakes access to the Atlantic Ocean. The Dutch settlers originally called the area "New Netherlands," the city "New Amsterdam." They built the wall across lower Manhattan for protection against the Indians and the British: "Wall Street" is now America's financial center. The famous Empire State Building has been topped twice by the towers of the World Trade Center (each one 1,350 ft., 412 m). Rural New York is a fertile farming region with a diverse landscape. At the Canadian border are the Niagara Falls, which are carving their way up the Niagara River toward Lake Erie at the rate of 4 ft. (1.2 m) per year.

## PENNSYLVANIA<sub>I</sub>

**Area:** 46,040 sq. mi. (119,151 km²). **Population:** 11,760,000. **Capital:** Harrisburg, 52,270. **Economy:** Manufacturing, iron and steel, coal, food products, dairying, and mushrooms. ☐ William Penn founded Pennsylvania as a religious sanctuary. It is called the "Keystone State" because of its central location within the original 13 colonies. Philadelphia (1,543,000), the largest city, was the nation's first capital and the economic, political, and cultural center of the colonies. It is still a major river port and a center for trade, business, culture, and education. The Pittsburgh area is the heart of a huge steel and coal industry. Some of the nation's richest farmland is found in the southeastern area, the Pennsylvania Dutch country, where Amish and Mennonite religious sects practice an early American life-style.

## RHODE ISLAND<sub>J</sub>

**Area:** 1,214 sq. mi. (3,142 km²). **Population:** 989,000. **Capital:** Providence, 156,900. **Economy:** Jewelry, silverware, textiles, and poultry (Rhode Island Red hens). ☐ Roger Williams was banished from Massachusetts in 1636 for being too liberal. He founded Rhode Island as a sanctuary for political and religious freedom—a policy that has remained firm through the years. The smallest state is no longer the textile center it once was. But Rhode Island is still an important manufacturing center, known for its jewelry and silverware. It has the highest proportion of industrial workers; only about 1% are involved in agriculture. Island-filled Narragansett Bay is a popular recreational area that occupies a large part of the state. Newport, at its mouth, is best known for yacht races, music festivals, and the most palatial summer estates in the nation. Providence has over 65% of the state's population.

## VERMONT<sub>K</sub>

**Area:** 9,610 sq. mi. (24,870 km²). **Population:** 560,000. **Capital:** Montpelier, 8,250. **Economy:** Timber products, precision manufacturing, maple syrup, granite, marble, asbestos, and dairying. ☐ Vermont, the least populous state east of the Mississippi, has the smallest percentage of city dwellers (30%). Burlington (39,600) is the largest city. Lakes and rivers define most of Vermont's boundaries; it is the only northeastern state without an Atlantic coastline. The scenic Green Mountains run through the center of the state. (The name "Vermont" comes from the French "vert mont," meaning "green mountain.") Stone quarries in these mountains provided most of the granite and marble used in building the nation's cities. Extensive forests supply many products, including the pure maple syrup for which Vermont is famous.

Long, hot summers, mild winters, heavy rainfall, and fertile land make the Southeast a major agricultural producer. Cotton, tobacco, and peanuts are the leading crops. The South, which also includes Texas, Louisiana, Arkansas, and Oklahoma, was once known as the "land of cotton." After World War II, the economic emphasis shifted to manufacturing, and northern companies came in search of cheap labor. New England's textile mills closed and moved closer to their source of cotton. The end of racial segregation removed a stigma associated with the South. The introduction of air conditioning further encouraged mass migration from the North. Only the West had more population growth. Even with industrialization and the arrival of newcomers, the South has retained much of its traditional charm.

The southeastern states have thick, fast-growing softwood forests that produce 40% of the nation's timber. With the exception of the low-lying Florida peninsula, the region slopes down from the Appalachian crests, across hilly plateaus (the piedmont) to the flat Atlantic coastal plain. From the mountains it also slopes northward to the Ohio River and westward to the Mississippi River. The rivers of the South, particularly the Mississippi and its tributaries, play a vital role in commerce, farming, and recreation. Almost all the large lakes in this region were created by the damming of rivers.

Washington, D.C. (District of Columbia), the nation's capital, is not part of any state. It has a local government, but it is under the control of Congress. Located on the Maryland side of the Potomac River (it was once part of Maryland), Washington is 69 sq. mi. (179 km²) in area and has a population of 638,500. The metropolitan Washington area is much larger than the District and includes parts of Maryland and Virginia.

## ALABAMA A

Area: 51,609 sq. mi. (133,667 km²). Population: 3,985,000. Capital: Montgomery, 197,100. Economy: Timber and paper, iron and steel, cotton, peanuts, pecans, soybeans, and textiles. □ The basic ingredients of steel production (iron, coal, and limestone) are mined locally, so Birmingham (286,800) has been the iron, steel, and heavy industry center of the South. The agricultural industry once revolved around cotton production. After an insect pest, the boll weevil, nearly wiped out an entire cotton harvest, farmers diversified. They now grow a wide variety of crops. This has led to increased prosperity and security—and the state erected a public monument in honor of the boll weevil. Montgomery was the first capital of the Confederacy. A hundred years later, Martin Luther King led a bus boycott there which accelerated the civil rights movement. The Marshall Space Flight Center in Huntsville has developed many of the rockets and missiles used in America's space program.

## FLORIDA B

Area: 58,664 sq. mi. (151,939 km²). Population: 12,800,000. Capital: Tallahassee, 81,600. Economy: Tourism, citrus, cattle, phosphates, vegetables, sugarcane, and electronics. □ Florida, a long (450 mi., 720 km), low, flat peninsula jutting into subtropical seas, is the one of the fastest growing states. The balmy climate makes year-round agriculture possible. The warm weather is attractive to both America's growing retirement community and the thousands of tourists who flock here. Popular attractions are Disney World in Orlando; the Kennedy Space Center at Cape Canaveral; St. Augustine, the first European settlement in the US (1565); the Florida Keys, a chain of coral islands; and Everglades National Park. The Everglades is a large, swampy region whose land and water are being devoured by the Florida building boom. Lake Okeechobee is the second largest fresh water lake (after Lake Michigan) completely within the US. Jacksonville (660,000) is physically the largest city in the country. South Florida is dominated by Miami (353,000) and its Latin American influence—largely the result of Cuban immigration. Florida has a huge citrus industry; nearly all frozen orange juice is processed there. Most of the country's phosphate, a chief ingredient of fertilizer, is mined in Florida.

## GEORGIA C

Area: 58,876 sq. mi. (152,489 km²). Population: 6,390,000. Capital: Atlanta, 385,000. Economy: Timber, aircraft, marble, cotton, tobacco, soybeans, peanuts, poultry, and peaches. □ Georgia was named after King George II. It is the largest state east of the Mississippi and one of the fastest growing. Atlanta, the capital, is the center of business, trade, and finance in the entire Southeast. Many varieties of trees, especially pines, cover 70% of the state. These pines produce half the world's supply of naval stores (tar, resin, and turpentine). Southeastern Georgia is a flatland covered by marshes and swamps (the Okefenokee is a lush wildlife refuge); bayous (narrow river outlets); and grassy plains called savannas. The city of Savannah is Georgia's second largest (147,000) and a seaport of historic importance. Undamaged by the Civil War, Savannah is a good example of antebellum architecture.

## KENTUCKY D

Area: 40,410 sq. mi. (104,662 km²). Population: 3,665,000. Capital: Frankfort, 25,980. Economy: Coal, tobacco, timber, bourbon whiskey, and racehorses. □ The name "Bluegrass State" comes from the blue tint of the grassy hills of the populated north-central region. Most of Kentucky is a hilly plateau that gradually slopes westward to the Mississippi. Along with the production of the best racehorses and bourbon whiskey, Kentucky is first in the US in coal production and second in tobacco products. The principal tourist attraction is Mammoth Cave, the largest system of caves (300

mi., 480 km) in the world. Louisville (298,750), the largest city, is an important inland port on the Ohio River and the home of the Kentucky Derby. Fort Knox is reputed to be the storehouse for all the gold owned by the US government.

## MISSISSIPPI E

Area: 47,718 sq. mi. (123,494 km²). Population: 2,535,000. Capital: Jackson, 202,980. Economy: Timber and paper, naval stores, cotton, soybeans, and oil. □ Mississippi was under French, British, and then Spanish rule before becoming a state. It was once one of the most prosperous states, but it took nearly 100 years to recover from the devastation it suffered in the Civil War. There was virtually no industry before World War II. Today, the state has one of the nation's most ambitious reforestation programs to protect its timber resources. Especially fertile farmland is found in alluvial regions (soil deposited by previous floods) along the Mississippi and also in the "Black Belt," a strip of dark, rich soil that runs across Mississippi and Alabama. The Gulf Coast is a popular resort area and home to the shrimp industry. Scenes of rural Mississippi life have been described in the writings of William Faulkner.

## NORTH CAROLINA F

Area: 52,586 sq. mi. (136,1988 km²). Population: 6,555,000. Capital: Raleigh, 150,640. Economy: Tobacco, textiles, furniture, timber, and cotton. □ North Carolina is the nation's largest producer of tobacco, textiles, and wooden furniture. In the western Blue Ridge Mountains is the tallest peak east of the Mississippi, Mt. Mitchell (6,684 ft., 2,038 m). The hilly, fertile piedmont region holds the major population and industrial centers as well as tobacco farms. To the east lie the savannas and swamps of the Atlantic coastal plain. Offshore is a chain of narrow islands, sandbars, and reefs, similar to those on Florida's east coast but far more treacherous. Shifting sands and unseen reefs have wrecked many a passing ship. Cape Hatteras has been called the "graveyard of the Atlantic." Kitty Hawk, a little to the north, is the stretch of sand made famous by the first flight of the Wright brothers in 1903.

## SOUTH CAROLINA G

Area: 31,112 sq. mi. (80,580 km²). Population: 3,408,000. Capital: Columbia, 102,000. Economy: Tobacco, textiles, paper products, and chemicals. □ The busy seaport of Charleston, the second largest city (69,550), was founded in 1670. Here one can still experience the charm and appearance of the "Old South." Aristocratic plantation owners lived a life-style patterned after that of English nobility. Their determination to retain their slave-based economy prompted South Carolina to become the first state to secede from the Union. The opening shots of the Civil War were fired on Ft. Sumter in Charleston harbor. Today, the state's rapidly expanding industrial economy is assisted by hydroelectric power, generated by 12 rivers rushing down from the "up country" in the west to the "low country" in the east. The central location of Columbia, the capital, was a compromise between rival settlers of these two "countries."

## TENNESSEE H

Area: 42,110 sq. mi. (108,981 km²). Population: 4,825,000. Capital: Nashville, 505,000. Economy: Chemicals, food products, vehicles, cotton, tobacco, and textiles. □ Tennessee is divided into three parts by the Tennessee River, which is unusual because it reverses direction after swinging south through Alabama. Along the river, the Tennessee Valley Authority (TVA) has erected 32 dams to create reservoirs and recreational lakes and provide hydroelectric power for Tennessee and seven neighboring states. Cotton and soybeans are the chief crops in the western region. Memphis, on bluffs above the Mississippi River, is the largest city (603,000) and busy inland port. Nashville, the capital and second largest city, is the country music capital of America (home of the Grand Ole Opry). Oak Ridge was the site of the nation's first nuclear reactor, which provided material for the first atomic bomb.

## VIRGINIA I

Area: 40,817 sq. mi. (105,716 km²). Population: 6,130,000. Capital: Richmond, 220,415. Economy: Chemicals, tobacco, tourism, shipbuilding, textiles, foods, and timber. □ Sir Walter Raleigh named Virginia for Queen Elizabeth I, the "Virgin Queen." Virginia's history dates back to the founding in 1607 of Jamestown, the first English settlement in North America. The state's tourist favorites are Williamsburg, a restored colonial village; Mt. Vernon and Monticello, the homes of George Washington and Thomas Jefferson (Virginia was also the birthplace of six other presidents); Arlington National Cemetery; the Blue Ridge Mountains; the Shenandoah Valley; and the many Civil War shrines (Virginia was the principal battleground). Proximity to the sea plays an important role in the state's strong economy. Norfolk, the largest city (267,215) and the major seaport, and Newport News are centers for shipbuilding and US naval installations.

## WEST VIRGINIA J

Area: 24,181 sq. mi. (62,629 km²). Population: 1,780,000. Capital: Charleston, 63,975. Economy: Coal, iron and steel, timber, chemicals, glassware, and marbles. □ Most of West Virginia's population and manufacturing centers are either on or close to the Ohio River. West Virginia is the most mountainous state in the nation. It has the highest average elevation of any state east of the Mississippi. The rugged terrain has bred a highly independent-minded citizenry. John Brown led his famous antislavery rebellion in 1859 at Harpers Ferry (now a national park). When the Civil War began, western Virginia residents, who had strong antislavery feelings, broke away from the plantation-owning easterners to form their own state, which then rejoined the Union. West Virginia's economy depends largely upon coal mining and industrial production. Indiscriminate strip mining and lumbering have ravaged many parts of this scenic region. Steps have been taken to restore the landscape.

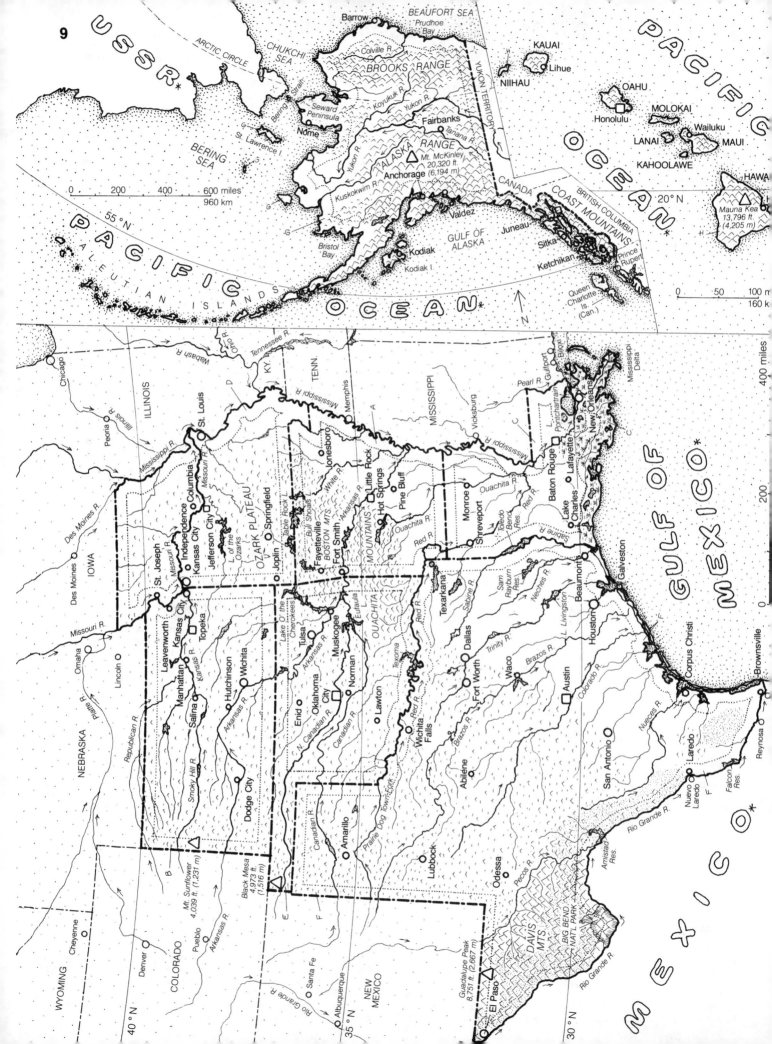

9

USSR*

ARCTIC CIRCLE

CHUKCHI SEA

BEAUFORT SEA
Barrow
Prudhoe Bay
Colville R.

KAUAI
Lihue

H

NIIHAU

BROOKS RANGE

YUKON TERRITORY

OAHU
Honolulu

MOLOKAI

Koyukuk R.

Bering Strait
Seward Peninsula

Yukon R.

Nome

St. Lawrence I.

Fairbanks
Tanana R.

Wailuku

LANAI

MAUI

BERING SEA

ALASKA RANGE
Mt. McKinley
20,320 ft.
(6,194 m)
Anchorage

CANADA

KAHOOLAWE

20°N

HAWAI

Kuskokwim R.

BRITISH COLUMBIA

Mauna Kea
13,796 ft.
(4,205 m)

G

55°N

Valdez

COAST MOUNTAINS

H

PACIFIC

Juneau

GULF OF ALASKA

Bristol Bay

Kodiak
Kodiak I.

Sitka

PACIFIC OCEAN*

0    200    400    600 miles
           960 km

ALEUTIAN ISLANDS

OCEAN*

Ketchikan

Prince Rupert

Queen Charlotte Is. (Can.)

N

0    50    100 m
        160 k

Chicago

ILLINOIS

Wabash R.

Ohio R.

Tennessee R.

KY.

TENN.

Mississippi R.

Memphis

MISSISSIPPI

Pearl R.

Gulfport
Biloxi

Mississippi Delta

Peoria
Illinois R.

St. Louis

D

A

Vicksburg

L. Pontchartrain

New Orleans

Mississippi R.

Missouri R.

Des Moines R.

Columbia
Missouri R.

Jonesboro

White R.

Little Rock

Pine Bluff

Ouachita R.

Monroe

C

Baton Rouge
Lafayette

Springfield

Joplin

Table Rock L.

Bull Shoals

BOSTON MTS.

Fort Smith

Hot Springs

Arkansas R.

Shreveport

Lake Charles

IOWA

Des Moines

St. Joseph
Independence
Columbia
Jefferson City

Kansas City

OZARK PLATEAU

L. of the Ozarks

Fayetteville

OUACHITA MOUNTAINS

Red R.

Toledo Bend Res.

Sabine R.

Beaumont

Galveston

GULF OF MEXICO*

Omaha

Lincoln

Leavenworth
Kansas City
Topeka

Manhattan

Salina

Kansas R.

Hutchinson

Lake O' the Cherokees

Tulsa

Muskogee

Eufaula L.

L. Texoma

Texarkana

Sam Rayburn Res.

Neches R.

L. Livingston

Houston

Corpus Christi

200

NEBRASKA

Platte R.

Republican R.

Smoky Hill R.

Mt. Sunflower
4,039 ft. (1,231 m)

Arkansas R.

Enid

Oklahoma City

Norman

Canadian R.

Lawton

Red R.

Dallas

Fort Worth

Waco

Brazos R.

Austin

Colorado R.

San Antonio

Laredo

Nueces R.

Brownsville

Reynosa

Missouri R.

Dodge City

Black Mesa
4,973 ft.
(1,516 m)

B

E

F

N. Canadian R.

Canadian R.

Prairie Dog Town Fork

Amarillo

Wichita Falls

Red R.

Abilene

Lubbock

Odessa

Pecos R.

Nuevo Laredo

Falcon Res.

Rio Grande R.

F

MEXICO*

WYOMING
Cheyenne

COLORADO

Denver

Pueblo

Arkansas R.

Santa Fe

Albuquerque

NEW MEXICO

Rio Grande R.

40°N

35°N

Guadalupe Peak
8,751 ft. (2,667 m)

El Paso

DAVIS MTS.

BIG BEND NATL PARK

Rio Grande R.

Amistad Res.

30°N

400 miles

# NORTH AMERICA: SOUTHCENTRAL US, ALASKA & HAWAII.

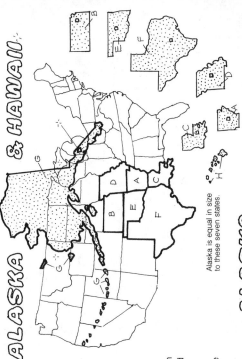

Of the six Southcentral states shown above, Kansas and Missouri are considered Midwestern; Texas, Louisiana, Arkansas, and Oklahoma, which fought for the Confederacy, are part of the South. Oklahoma was an Indian Territory during the Civil War, but a number of its tribes fought on the Southern side.

Oil has been the driving force behind the economies of Texas, Oklahoma, Louisiana, and natural gas deposits in this region have gained importance. Alaska also became a major oil producer with the discovery of large oil reserves on its northern coast. But for all these states boom turned to depression in the 1980s when the worldwide price of oil fell sharply.

## ARKANSAS A

**Area:** 53,103 sq. mi. (137,537 km²). **Population:** 2,338,000. **Capital:** Little Rock, 159,100. **Economy:** Timber, rice, broiler chickens, oil and natural gas, and bauxite. □ Arkansas (ark' in saw) is best known for its scenery and outdoor attractions, especially its natural springs. The therapeutic waters of over 40 springs in the city of Hot Springs draw millions of visitors each year. The springs are actually in Hot Springs National Park, most of which is within city limits. Arkansas leads the nation in rice and broiler chicken production. The state has America's only diamond mine. Visitors may keep whatever diamonds they find. The eastern part of the state, along the Mississippi, has fertile alluvial soil. Rich farmland can also be found in the northern Ozark region. Arkansas has one of the highest rural populations in the nation.

## KANSAS B

**Area:** 82,264 sq. mi. (212,899 km²). **Population:** 2,468,000. **Capital:** Topeka, 115,800. **Economy:** Wheat, cattle, oil and natural gas, and aircraft. □ The nation's leading wheat producer has a colorful western past—cattle drives, gunfighters, saloons, and wagon trails. Railroad terminals in Abilene and Dodge City were the destination for many cattle drives coming from as far away as Texas. The geographical center of the 48 states is in central Kansas, near the Nebraska border. The major population centers are in the eastern half of the state, which has rich alluvial soil and adequate rainfall. In the drier west, there are very few trees, and the land is a flat plain gradually ascending to 4,000 ft. (1,220 m) to the foothills of the Rockies. Land conservation, irrigation, dams, and reservoirs have revived Kansas from the "dust bowl" condition of the 1930s. Wichita, the largest city (280,100), is the nation's number one producer of private aircraft.

## LOUISIANA C

**Area:** 48,522 sq. mi. (125,575 km²). **Population:** 4,180,000. **Capital:** Baton Rouge, 220,500. **Economy:** Oil and natural gas, chemicals, salt, sulphur, soybeans, sugar, and shrimp. □ The huge Delta region was formed by silt (particles of soil suspended in water) donated by states located up the Mississippi River. A massive network of dikes and levees (banks) has been built to protect against flooding. The state's extensive Gulf coastline includes hundreds of inlets, islands, sandbars, bayous, and marshes. This area supports huge pelican and egret populations and is the winter home for half the wild ducks and geese in North America. New Orleans, the largest city (487,000), lies 100 miles upriver from the Gulf. It is a major international shipping center and one of America's favorite tourist attractions with its fine restaurants, Old World architecture, street art, Dixieland music, and the Mardi Gras. Louisiana, more than any other state, has a distinctly foreign flavor. The civil law is based upon France's Napoleonic Code. Many people in the southern half of the state are either Cajuns or Creoles, and most of them are Roman Catholic. Cajuns, who speak a French dialect as well as English, are descendants of early Canadian settlers from Nova Scotia. Creoles are people of French or Spanish ancestry. Northern Louisiana was primarily settled by Anglo-Saxon Protestants from other states.

## MISSOURI D

**Area:** 69,685 sq. mi. (180,345 km²). **Population:** 5,080,000. **Capital:** Jefferson City, 33,680. **Economy:** Transportation equipment, lead, soybeans, corn, and meat-packing. □ Because of its central location and access to the nation's largest rivers, Missouri (muh zoor' ee) is a major trade and transportation center. In the 19th century, the Sante Fe and Oregon Trails led west from the town of Independence. The Pony Express, which ran to California, started from St. Joseph. Close to the Mississippi riverfront in St. Louis (393,000) is the nation's tallest monument, the 630 ft. (192 m) Gateway Arch. It is dedicated to the city's historic role as the gateway to the West. St. Louis and Kansas City (427,000) are also major air, rail, and trucking centers, as well as important inland ports. Missouri also produces transportation equipment: autos, planes, railroad cars, buses, and aerospace products. The hilly Ozark Plateau region is a popular recreation area and the center for the nation's largest lead mining industry.

## OKLAHOMA E

**Area:** 69,920 sq. mi. (180,953 km²). **Population:** 3,124,000. **Capital:** Oklahoma City, 404,600. **Economy:** Oil and natural gas, cattle, and wheat. □ "Oklahoma" is an Indian term meaning "land of red people." But with the firing of a single pistol shot in 1889, which signaled the start of a massive land grab by white settlers, the Indian population lost most of their treaty rights. Those settlers, who raced out to claim their land before the gun was fired were called "Sooners"; the nickname now applies to all Oklahomans. In Oklahoma, oil derricks can be seen everywhere, including the front lawn of the state capitol. Tulsa (365,000), the second largest city, has been called the "oil capital of the world." When an enormous navigation project on the Arkansas River made possible commercial traffic to the Mississippi, Tulsa became an important inland port. Grasslands support a huge beef industry, but the soil has never recovered its pre-dust bowl fertility. The effects of the droughts of the 1930s and the emigration of thousands of "Okies" from the state were chronicled by John Steinbeck in *The Grapes of Wrath*.

## TEXAS F

**Area:** 267,338 sq. mi. (691,872 km²). **Population:** 16,830,000. **Capital:** Austin, 462,000. **Economy:** Oil and natural gas, cattle, cotton, sulphur, machinery, electronics, and machinery. □ The rallying cry "remember the Alamo" led a Texas army to victory and independence from Mexico in 1836. The Alamo was a San Antonio chapel in which Texas revolutionaries (including Jim Bowie and Davy Crockett) were wiped out by a Mexican army. Alaska displaced Texas as the largest state, but Texans can still boast of being number one in many other departments. Texas produces the most oil, natural gas, sulphur, asphalt, and gypsum. Texas also has the most ranches and farms, which raise the most cattle, horses, sheep, wool, and cotton. The King Ranch alone is larger than Rhode Island. Texas ranks third, after California and New York, in the value of the goods it produces. Houston (1,610,000), a major seaport, is the trade and manufacturing center for this region of the country. The Johnson Space Center is located close by. Dallas (990,000) is the heart of the oil, banking, and insurance industries. Most of Texas's business and population centers are located in this eastern region.

## ALASKA G

**Area:** 586,400 sq. mi. (1,517,603 km²). **Population:** 550,000. **Capital:** Juneau, 19,600. **Economy:** Oil, natural gas, timber, fishing, and gold. □ Alaska means "great land" in Aleut—and it surely lives up to its name. It is more than twice the size of Texas; it has a longer coastline than all other states combined; and it has the tallest mountain peak in North America (Mt. McKinley). In 1867, Secretary of State Seward pressured Congress to buy "Russian America" for $7.2 million. Even at 2 cents an acre (5 cents a hectare), it was referred to as "Seward's Folly." The US and the USSR are separated by the 50 mi. (80 km) wide Bering Strait. Alaska, America's last frontier, is a mostly uninhabited region with a broad range of untapped mineral and natural resources. Its fishing industry (principally salmon) is the nation's largest. The winters are generally frigid, except in the southern and southeastern regions, which have temperatures comparable to many northern US cities. Barrow, on the barren Arctic coast, is the largest Eskimo village in the world. Oil was discovered in Prudhoe Bay and a pipeline was built to transport it across the state to Valdez, a small town on Prince William Sound in the Gulf of Alaska, where a devastating oil spill occurred in 1989. Alaska's worst weather and most hostile landscape is in the Aleutian Islands, an archipelago extending westward for 1,700 mi. (2,720 km). Japanese troops occupied the two outermost islands during World War II.

## HAWAII H

**Area:** 6,450 sq. mi. (16,706 km²). **Population:** 1,100,000. **Capital:** Honolulu, 377,000. **Economy:** Military installations, tourism, pineapples, and sugar. □ Entirely created by volcanoes, Hawaii (huh wy' ee) is a 1,500 mi. (2,400 km) archipelago of 130 islands in the mid-Pacific, southeast of the mainland and about the same distance from San Francisco as is New York City. The islands were settled by Polynesians from Southeast Asia around 750 AD and were discovered in 1778 by the English explorer James Cook, who named them the Sandwich Islands. Hawaii has the broadest ethnic mix of any state and the highest percentage of Asians (58%). Most Hawaiians live on the five largest islands shown above. Hawaii followed Alaska in entering the Union in 1959. Beautiful weather, a lush tropical environment, native hospitality, and jet travel have created a thriving tourist trade. Honolulu, on the island of Oahu, is the business center and has over 80% of the state's population. Also located on Oahu is Pearl Harbor, the seaport and naval base that was attacked by the Japanese, bringing the US into World War II. Hawaii, the "Big Island," was formed by five volcanoes; two of them are still active. Maui has the world's largest inactive crater, about 7 mi. (11.2 km) across. A mountaintop on Kauai, the greenest island, is the wettest spot on the planet, with 460 in. (1,168 cm) of rain annually.

Alaska is equal in size to these seven states.

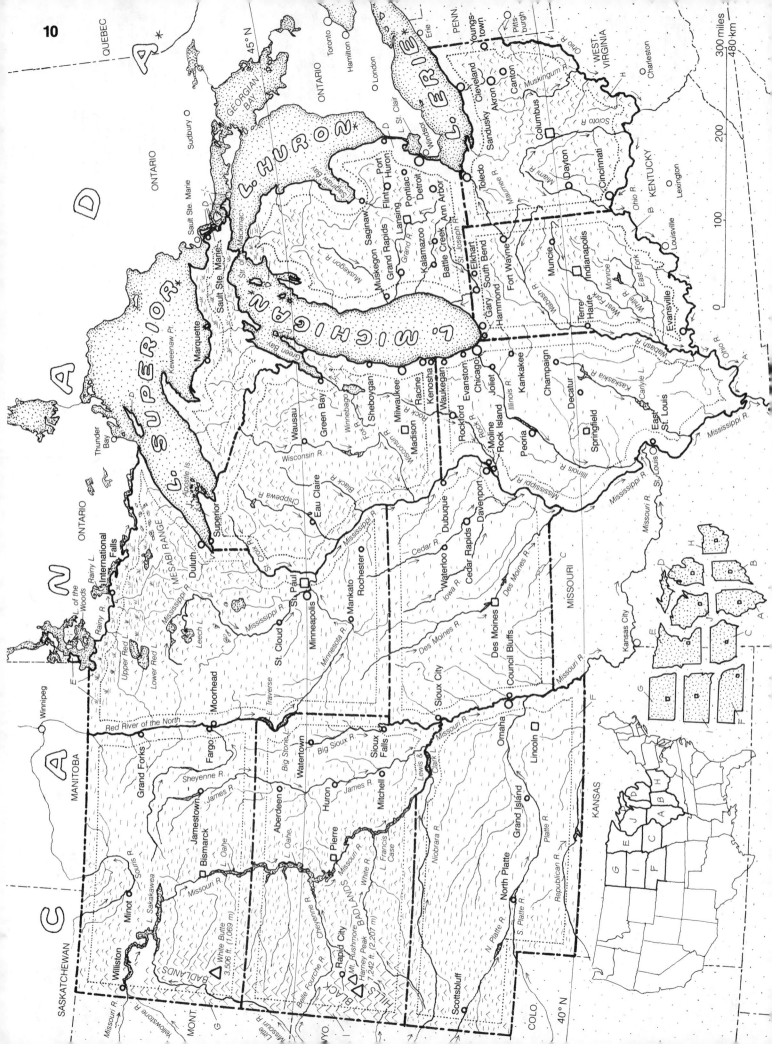

# NORTH AMERICA: NORTHCENTRAL US ₁₁

These states, along with Kansas and Missouri (Plate 9), make up the American Midwest. This is one of the world's most productive agricultural areas. Its central location gives it access to transportation on major rivers (the Mississippi, Missouri, and Ohio), the Great Lakes and St. Lawrence Seaway, and extensive rail and road networks. The lower tier of states makes up the "corn belt." Much of America's most valuable crop is grown to feed cattle, which are sent to feedlots in the Midwest for fattening before slaughter. Wheat, the most important food crop, dominates the western fringe of states. The northernmost states specialize in "spring wheat" while those to the south grow "winter wheat." In addition to the two wheat belts, there is a "dairy belt" across the northcentral states; Wisconsin is the leading dairy producer. The five states east of the Mississippi also became industrial powerhouses as well as major agricultural producers. These states were ideally located between the iron deposits of northern Minnesota, Wisconsin, and Michigan and the coal deposits of the Appalachian Mountains. Interstate commerce via the Great Lakes was expanded to international trade when the completion of the St. Lawrence Seaway, in Canada, opened a route to the Atlantic (see map on Plate 6). Except for the northern part of Minnesota, Wisconsin, and Michigan, this is essentially a treeless region. Much of the land is rolling hills or flat prairie that slowly climbs to a height of 5,000 ft. (1,524 m) at its western edge.

## ILLINOIS ₐ

**Area:** 56,400 sq. mi. (145,963 km²). **Population:** 11,325,000. **Capital:** Springfield. 100,125. **Economy:** Corn, soybeans, heavy machinery, food products, hogs, coal, and electronics. ☐ As a leading agricultural and industrial state, Illinois (ill un noy) is number one in corn, soybeans, meat-packing, farm and road-building equipment, and diesel engines. Chicago (2,725,000), the nation's transportation hub, handles more passengers and freight through its air, ship, rail, truck, and bus terminals than any other city in North America. With the opening of the St. Lawrence Seaway, Chicago became the world's busiest inland port. It is the trade, financial, and cultural center of the Midwest. It has the world's busiest airport, O'Hare, and the tallest building, the 110-story Sears Tower (1,454 ft., 443 m). Over 60% of the state's population lives in the greater Chicago area. In the suburb of Batavia, the Fermi National Accelerator Lab is a leader in atomic research. Illinois, Ohio, and Wabash Rivers; Lake Michigan and the Mississippi, Illinois, Ohio, and Wabash Rivers.

## INDIANA ₆

**Area:** 36,280 sq. mi. (93,893 km²). **Population:** 5,499,000. **Capital:** Indianapolis. 737,000. **Economy:** Coal, iron and steel, electrical and food products, corn, soybeans, hogs, and limestone. ☐ Though only 38th in size, Indiana ranks among the top agricultural and industrial states. The Lake Michigan riverfront of Gary and Hammond extends 50 mi. (80 km); it is one of the world's great industrial regions—a center for iron and steel, chemicals, oil refineries, and general manufacturing. Indianapolis, the capital and business center, is a major livestock distribution hub because many interstate truck routes pass through it. The city is also the home of the Memorial Day automobile racing classic, the Indianapolis 500. Notre Dame University is located in South Bend. Nearby Elkhart is the leading producer of band and orchestral instruments. Most of the limestone used in the nation's building industry comes from quarries in southern Indiana.

## IOWA ꜀

**Area:** 56,274 sq. mi. (145,693 km²). **Population:** 2,766,000. **Capital:** Des Moines. 197,850. **Economy:** Corn, hogs, soybeans, food products, and farm machinery. ☐ The golden, gently rolling hills made famous in the paintings of Grant Wood hint at the agricultural richness of this state. Iowa has as much as 25% of the nation's best farmland. Though it is medium-sized in area and population, Iowa ranks second only to California in food production. Iowa alternates with Illinois as the leading corn producer. It is first in hogs, second in soybeans, and one of the leaders in cattle, most of which are brought into the state to be "corn-fattened" before slaughter. A quarter of Iowa's food production is exported abroad. Des Moines is the capital and largest city.

## MICHIGAN ᴅ

**Area:** 58,220 sq. mi. (150,731 km²). **Population:** 9,179,000. **Capital:** Lansing. 130,420. **Economy:** Autos, machinery, iron ore, oil, food processing, salt, and tourism. ☐ Though far from any ocean, Michigan is almost all coastline. The state consists of two peninsulas, surrounded by four of the five Great Lakes (all but Lake Ontario). Most of the population, factories, and farms are located in the lower half of the larger peninsula. Only 3% live in the iron-rich, upper peninsula. At its tip is one of the busiest canals in the world, Sault Ste. Marie, which connects Lake Superior with Lake Huron. The forests of the north, along with over 10,000 lakes and the extensive coastline, make Michigan a leading recreational area. More cars and trucks are produced in Detroit (970,000), Flint, Lansing, and Pontiac than any comparable region in the world. In the processing of grains, Battle Creek is synonymous with breakfast cereal. Much of the nation's freshwater fish comes from Michigan's many inland fisheries.

## MINNESOTA ₑ

**Area:** 84,401 sq. mi. (218,514 km²). **Population:** 4,360,000. **Capital:** St. Paul, 271,235. **Economy:** Dairying, grains, iron ore, timber, and electronics. ☐ The Indian name "Minnesota" means "sky-blue water," a fitting title for a land of rivers, streams, swamps, waterfalls, and over 15,000 lakes. Glacial action flattened large sections of the state but was not as effective in the hilly, iron-rich northeastern "superior uplands" region—the lower tip of the rocky Canadian Shield. There, in the Mesabi Range, the largest open-pit mine in the world has been extracting iron for 80 years. The premium ore was exhausted, but newer technology has made it feasible to continue mining inferior grades, now 60% of the nation's production. The ore is shipped, along with timber and grains, out of Duluth and Superior, Wisconsin, the westernmost ports on the Great Lakes. Half the population (including the most Scandinavians in the US) lives in the twin cities of Minneapolis (367,000) and St. Paul, modern centers of business and culture. The famed Mayo Clinic is located in Rochester. Because it is a leading grain and dairy producer, Minnesota is often called the "bread and butter state."

## NEBRASKA ꜰ

**Area:** 77,354 sq. mi. (200,270 km²). **Population:** 1,573,000. **Capital:** Lincoln, 171,992. **Economy:** Wheat, corn, cattle, hogs, food products, alfalfa, and oil. ☐ The "Cornhusker State" is actually third in corn production but first in alfalfa and second only to Texas in cattle. As much as 95% of its land is devoted to agriculture. The corn-growing farms are located in the wetter eastern half of the state. Wheat and cattle are raised in the west. A greater variety of grass is grown for forage in most parts of the state than anywhere else. Omaha (333,000), the largest city, is a major livestock trading center and the hub of finance and insurance. Nebraska is the only state with a single-house legislature whose members are elected without any party affiliation. In the 19th century, many Nebraskans provided services for pioneers passing through the state on their way west. The famous Oregon Trail ran along the Platte and North Platte Rivers.

## NORTH DAKOTA ꜱ

**Area:** 70,700 sq. mi. (183,042 km²). **Population:** 634,000. **Capital:** Bismarck, 44,490. **Economy:** Oil, lignite coal, wheat, flax, sunflower seeds, barley, and oats. ☐ In this sparsely populated state, only four cities have over 25,000 residents. Fargo (61,500) is the largest. North Dakota is the nation's leading producer of lignite (a soft coal); spring and durum wheat (the latter used in pasta); flax seed (pressed into linseed oil); sunflower seeds; and barley. Coal and oil deposits are mined in the rugged western region. The narrow Red River Valley, on the eastern border, is a remarkably fertile ancient lake bed. Here, in the 19th century, large wheat farms were so profitable they were called "bonanza" farms. Immigration was encouraged, and in the 19th century many Norwegians and Germans settled in this valley. The Garrison Dam on the Missouri created Lake Sakakawea, 178 mi. (285 km) long.

## OHIO ₕ

**Area:** 41,228 sq. mi. (106,739 km²). **Population:** 10,777,000. **Capital:** Columbus, 565,250. **Economy:** Machinery, aircraft parts, iron and steel, coal, rubber products, corn, and soybeans. ☐ Despite its modest size, Ohio's strategic location, coal reserves, and fertile soil have made it a major industrial and agricultural state. It is the top producer of rubber products and machine tools and is a leader in iron and steel, aircraft parts, general manufacturing, corn, soybeans, and hogs. The "Buckeye State" was covered with buckeye trees before the land was cleared for farming. Cleveland (500,000) is no longer larger than Columbus (625,000), the capital. Cincinnati (352,000) is on the Ohio River, a major tributary of the Mississippi.

## SOUTH DAKOTA ₁

**Area:** 77,114 sq. mi. (199,648 km²). **Population:** 694,000. **Capital:** Pierre, 11,976. **Economy:** Food processing, tourism, wheat, cattle, sheep, and gold. ☐ After agriculture, tourism is the second largest industry. Visitors come to see the Black Hills (highest mountains east of the Rockies); the Badlands (grotesque formations of rock and clay in a desert environment); and the four presidential heads (Washington, Lincoln, Jefferson, and Teddy Roosevelt) carved in the granite face of Mt. Rushmore. Also of interest is Deadwood, once the most lawless town on the frontier, the haven for gold prospectors, gamblers, dancehall girls, and gunslingers. Gold is still mined in the Black Hills at the largest mine in North America. South Dakota has a long history of injustice and violence against the native people, particularly the Sioux. Their descendants have received reparations from Congress for land taken from them in the 19th century. The state is sparsely settled: Sioux Falls (82,100) is the largest of only three cities to exceed 25,000. The Missouri bisects the state into a rugged western part and a flat, fertile eastern half. The construction of four large dams on the Missouri in the 1930s created a major source of irrigation and hydroelectric power.

## WISCONSIN ⱼ

**Area:** 56,155 sq. mi. (145,386 km²). **Population:** 4,870,000. **Capital:** Madison, 171,750. **Economy:** Dairying, engines and turbines, food and paper products, and beer. ☐ Wisconsin is called America's dairy (it produces 40% of the nation's cheese) but the state's economic wealth is in industry: paper products, engines for outboards and lawnmowers, beer brewing, and food processing. The northern region, with 15,000 lakes, rivers, and waterfalls, is a popular recreational area. Despite the indiscriminate lumbering of the past, a highly successful program has restored much of Wisconsin's forest, which covers half the state. Milwaukee (620,000), the largest city and industrial center, is the beer-brewing capital of the country—and not surprisingly, home to the nation's largest German-American community. A German creation, the kindergarten, was introduced to America over 100 years ago in a small Wisconsin village. Over the years, Wisconsin has led the way in passing progressive legislation regarding jobs, welfare, health, and public safety.

Except for the few states whose economies rise or fall with the price of oil, the West is the fastest growing part of America. Since World War II there has been a steady shift in population from the Northeast and Midwest to the West and the South. Except for the Pacific Northwest, these states are virtual deserts that depend on irrigation provided by rivers and reservoirs. The West has the nation's tallest mountains, highest plateaus, deepest canyons, wettest rain forests, and driest deserts. The region is bracketed by the Coast Ranges, Cascades, and Sierra Nevada Mountains to the west and the massive Rocky Mountains to the east. Between them lie the high basins and plateaus sometimes referred to as the Intermountain Area. These flat lands are partially covered by smaller mountain ranges.

## ARIZONA A

**Area:** 113,950 sq. mi. (295,130 km²). **Population:** 3,625,000. **Capital:** Phoenix, 972,100. **Economy:** Electronics, manufacturing, copper, metals, cotton, cattle, and tourism. □ Reservoirs, irrigation, and air conditioning have transformed a barren desert into a booming industrial and agricultural economy. Most of the population lives and works in the hotter and drier southern half of the state. Tucson (403,000) and Phoenix are agricultural and industrial centers. Arizona's Hopi, Navajo, and Apache tribes make up the nation's largest Indian population (120,000). Near Phoenix is America's first apartment house, Casa Grande, an 800-year-old, four-story adobe (sun-dried brick) structure. The majestic Grand Canyon is the product of 6 million years of erosion by the Colorado River with help from rain and snow runoff. The rocks at the bottom of the nearly 1 mi. (1.6 km) deep canyon are 2 billion years old.

## CALIFORNIA B

**Area:** 158,700 sq. mi. (411,033 km²). **Population:** 29,500,000. **Capital:** Sacramento, 365,000. **Economy:** Aircraft, space equipment, electronics, oil, produce, and cotton. □ The manufacturing and agricultural output of the most populous state rank it as the world's sixth largest economy. The mild, Mediterranean climate permits year-round agriculture in the wide, fertile, irrigated Central Valley, which is 500 mi. (800 km) long. California is the leading producer of fruits, nuts, vegetables, cotton, and flowers. It is also the leading industrial state. Los Angeles (3,425,000) is a sunny, colorful, flat, sprawling city plagued by smog. It is the nation's second largest city and top manufacturing center. The part of it called Hollywood is the entertainment capital of the world. San Francisco (712,000) is the opposite of L.A.: confined to a small peninsula, it has breathtaking hills, cable cars, Victorian houses, chilly fog, and the Golden Gate Bridge. The fastest growing cities are San Diego (1,110,000) and San Jose (766,000). California's landscape is remarkably diverse: a dramatic coastline, snowcapped mountains, fertile valleys, thick forests, and barren deserts. The tallest peak in the 48 states is Mt. Whitney (14,494 ft., 4,418 m)—60 miles away is Death Valley, the lowest and hottest place in North America (-282 ft., -86 m). California's redwoods are the world's tallest trees; the sequoias are the largest; and the bristle-cone pines are the oldest (4,000 years).

## COLORADO C

**Area:** 104,200 sq. mi. (269,878 km²). **Population:** 3,274,000. **Capital:** Denver, 460,000. **Economy:** Oil, coal, precision manufacturing, minerals, cattle, and tourism. □ With an average elevation of 6,800 ft. (2,073 m), Colorado is the highest state. It has over half of the 54 Rocky Mountain peaks taller than 14,000 ft. (4,268 m). The headwaters of the Colorado and Rio Grande Rivers and tributaries of the Missouri originate in these mountains. Manufacturing, agriculture, and tourism have replaced mineral wealth as the basis for the state's expanding economy. Colorado's mountains have enormous deposits of shale oil. The population centers are on the dry eastern slopes of the Rockies; water is brought through mountain tunnels in order to tap the greater runoff on the western side of the Great Divide. Colorado Springs is the site of the Air Force Academy, and buried deep within a nearby mountain is the headquarters for the North American Air Defense Command.

## IDAHO D

**Area:** 83,560 sq. mi. (216,420 km²). **Population:** 1,005,000. **Capital:** Boise, 103,300. **Economy:** Potatoes, timber, silver, food processing, and minerals. □ Scenic beauty abounds in this sparsely populated, mountainous state, whose northern panhandle has been designated a wilderness preserve. Tourists are attracted to the thousands of lakes and streams; caverns; high waterfalls; and steep canyons (Hells Canyon, on the Snake River, is deeper than the Grand Canyon). Most cities and farms are located on the highly irrigated Snake River Plain in the south, where the nation's largest potato crop is grown and the largest Mormon community outside Utah is found. A massive navigation project along the Snake River has opened a sea route from Idaho to the Pacific Ocean via the Columbia River. Idaho is the leading producer of silver, phosphate rock, and molybdenum (used in hardening steel).

## MONTANA E

**Area:** 147,250 sq. mi. (381,377 km²). **Population:** 795,000. **Capital:** Helena, 23,940. **Economy:** Oil, coal, minerals, wheat, cattle. □ Montana means "mountain" in Spanish, but three-fifths of the state lies in the eastern high plains, where wheat, cattle, and sheep are the dominant industries. The state's most dramatic scenery is at the Waterton-Glacier International Peace Park on the Canadian border, where some 60 glaciers are on the move. The earliest settlers came in search of gold and silver— the primary reason many people came West. Most of the smaller cities in this thinly populated state began as Rocky Mountain mining towns. Billings (66,875) and Great Falls (56,900) are located in the high plains. Rich in gold, silver, and precious stones, Montana has been called the "Treasure State." The "cowboy and Indian" past is evident in the popularity of rodeos and traditional Indian ceremonies.

## NEVADA F

**Area:** 110,550 sq. mi. (286,325 km²). **Population:** 1,200,000. **Capital:** Carson City, 32,100. **Economy:** Tourism and gambling, gold, minerals, and manufacturing. □ The Sierra Nevada Mountains screen out Pacific-bred rainstorms, so Nevada, the driest state, is totally dependent on irrigation projects. Millions of visitors are attracted by unrestricted gambling, big-name entertainment, legalized prostitution, and liberal divorce and marriage laws. A string of ghost towns is a reminder of a 19th century rush for gold and silver. The land is a high basin, averaging 5,000 ft. (1,524 m), with rows of mountain ranges running north and south. The US owns 87% of the state and operates test centers here for nuclear energy and weapons

## NEW MEXICO G

**Area:** 121,620 sq. mi. (314,996 km²). **Population:** 1,492,000. **Capital:** Santa Fe, 49,400. **Economy:** Oil, natural gas, coal, uranium, electronics, cattle, sheep, and tourism. □ This ruggedly beautiful state has a long history: stone age civilization, centuries of pueblo (village) dwellers, Spanish occupation, Mexican rule, Confederate occupation, and US territorial status before achieving statehood. Santa Fe, the capital, has the nation's oldest government building, first used by the Spanish in 1609. Sante Fe's adobe architecture is unique among American cities. The first road in what is now the US was built in 1581; it ran from Sante Fe to Mexico City. The US government conducts space and nuclear energy operations around Albuquerque (336,400), the largest city and manufacturing center. The state has huge uranium deposits. The first atomic bomb was created in Los Alamos and exploded near Alamogordo.

## OREGON H

**Area:** 97,040 sq. mi. (251,337 km²). **Population:** 2,830,000. **Capital:** Salem, 89,550. **Economy:** Timber products, wheat, food products, and electronics. □ Two distinct climates characterize Oregon. West of the towering Cascades, the weather is mild and moist. In this region lie Oregon's vast forests and the fertile Willamette River Valley, with its major cities, industries, and productive farmland. Oregon is the nation's leading timber state. The eastern two-thirds of Oregon consists of a dry plateau subject to wide variations in temperature. Irrigation has made this a productive agricultural area. Portland (435,000), the largest city, is an important port on the Columbia. This river provides much of the Northwest's hydroelectric power. Oregon has many scenic attractions: Crater Lake, the country's deepest lake, which fills the crater of an extinct volcano; steep gorges on the Columbia and Snake Rivers; Pacific beaches; and the majestic Mt. Hood.

## UTAH I

**Area:** 84,912 sq. mi. (219,922 km²). **Population:** 1,712,000. **Capital:** Salt Lake City, 163,500. **Economy:** Oil, coal, heavy equipment, electronics, and minerals. □ Utah is either desert or mountain. The state's wealth is in mineral deposits. Less than 10% of the land is arable, and only half of that is under cultivation. The western third, in the dry Great Basin, was the ancient Lake Bonneville. This area holds rock-hard salt flats and the Great Salt Lake, 6 times saltier than the ocean. (Without outlets, fresh water flowing into the lake evaporates, leaving salt residues.) The capital, Salt Lake City, is the headquarters of the Mormon Church, whose members make up 70% of the state's population. The city was founded in 1847 by the Mormon leader, Brigham Young. Utah's statehood was delayed by Congress for nearly 50 years until the church banned polygamy. Near the Arizona border are two natural wonders: Monument Valley, remarkable for giant red sandstone formations that rise 1,000 ft. (305 m) from the flat desert floor; and Bryce Canyon, noted for the brilliant colors and bizarre shapes of its huge eroded rock structures.

## WASHINGTON J

**Area:** 68,190 sq. mi. (176,612 km²). **Population:** 4,830,000. **Capital:** Olympia, 27,540. **Economy:** Timber, aircraft, shipbuilding, fruit, and fishing. □ The Puget Sound region has hundreds of islands, natural harbors, forests, two mountain ranges, and Mt. Rainier. A volcanic peak in the Cascades, Mt. St. Helens, has had several eruptions since 1980. West of Puget Sound are the wet and primitive Olympic Mountains. Seattle (514,000), the largest city, is just one of many Washington locations for the air and space giant, the Boeing Company. East of the Cascades is a large plateau formed by ancient lava deposits. This arid region has been made into a productive farming area with water and electricity supplied by dams on the Columbia River. The Grand Coulee is the largest cement dam in the US. Washington grows more apples and hops (used in brewing beer) than does any other state.

## WYOMING K

**Area:** 97,904 sq. mi. (253,571 km²). **Population:** 449,000. **Capital:** Cheyenne, 47,250. **Economy:** Oil, natural gas, coal, uranium, cattle, and sheep. □ The ninth largest state has the fewest people. Sheep and cattle raising are major industries. The state is second only to Colorado in average elevation. Despite its "wild west" reputation, Wyoming is called the "Equality State": it was the first to grant women the rights to vote, to hold public office, and to serve on juries. It even had the first woman governor, who in 1925 was allowed to finish out her deceased husband's term. Two of America's most beautiful national parks are in the Rocky Mountain region: Yellowstone, the oldest and largest, and Grand Teton. Casper (51,000), is the largest city.

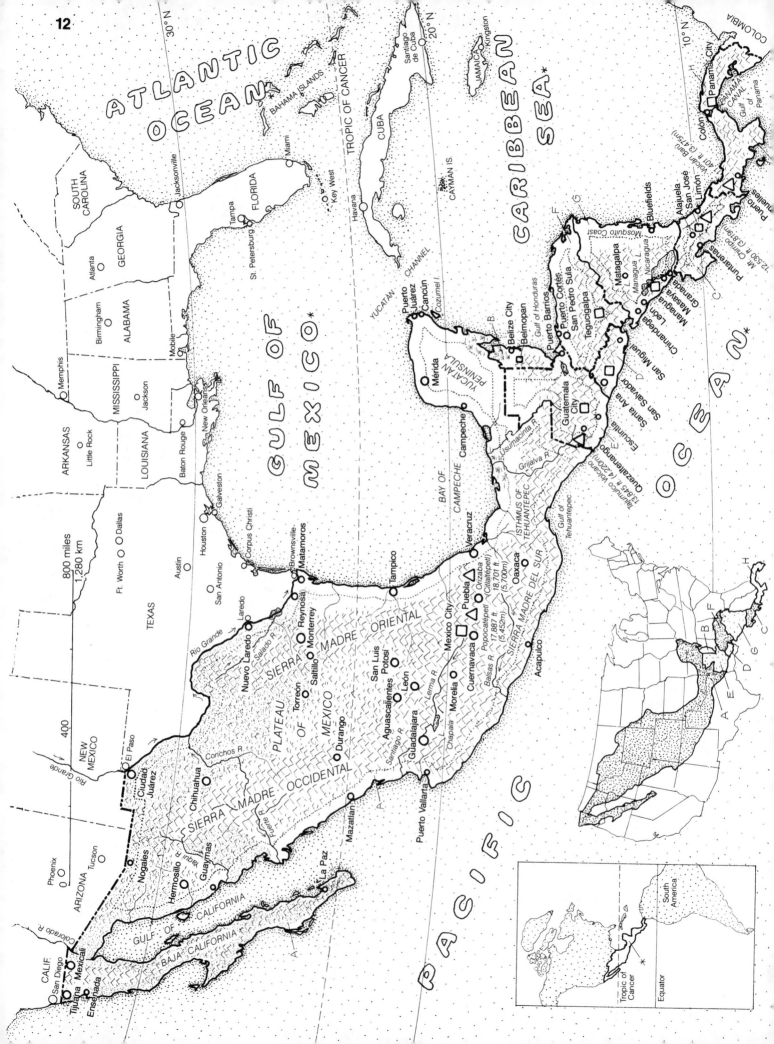

ATLANTIC OCEAN *

SOUTH CAROLINA

GEORGIA

ALABAMA

MISSISSIPPI

ARKANSAS

LOUISIANA

TEXAS

NEW MEXICO

ARIZONA

CALIF.

Atlanta
Birmingham
Memphis
Jackson
Little Rock
Baton Rouge
New Orleans
Mobile

Jacksonville
Tampa
St. Petersburg
Miami
Key West
FLORIDA

BAHAMA ISLANDS

TROPIC OF CANCER

30° N
20° N
10° N

CUBA
Havana
Santiago de Cuba
Kingston
JAMAICA
CAYMAN IS.

YUCATAN CHANNEL

CARIBBEAN SEA *

COLOMBIA

GULF OF MEXICO *

BAY OF CAMPECHE

YUCATAN PENINSULA

Mérida
Campeche
Cozumel I.
Cancún
Puerto Juárez

Belize City
Belmopan
BELIZE
Puerto Barrios
Gulf of Honduras
Puerto Cortés
San Pedro Sula
Tegucigalpa

HONDURAS

Mosquito Coast
Bluefields

NICARAGUA
Managua L.
Nicaragua L.
Matagalpa
León
Managua
Masaya
Granada
Chinandega

COSTA RICA
Alajuela
San José
Limón
Puntarenas

PANAMA
Panama City
Colón
PANAMA CANAL
Gulf of Panama
Puerto Armuelles
Volcán Barú (3,475m)
11,401 ft.

Mt. Chirripó 12,530 ft. (3,819m)

Guatemala City
GUATEMALA
Quezaltenango
Tajumulco Volcano 13,845 ft. (4,220m)
Escuintla
Santa Ana
Usumacinta R.
Grijalva R.
San Miguel
EL SALVADOR
San Salvador

San Diego
Tijuana
Mexicali
Ensenada

Nogales
Tucson
Phoenix
Hermosillo
Guaymas

BAJA CALIFORNIA
GULF OF CALIFORNIA
La Paz

SIERRA MADRE OCCIDENTAL

Ciudad Juárez
El Paso
Rio Grande
Colorado R.

Chihuahua
Conchos R.
Fuerte R.
Mazatlán
Durango
Torreón
Saltillo
Monterrey
Reynosa
Nuevo Laredo
Laredo
Salado R.
Brownsville
Matamoros
Corpus Christi
Houston
Galveston
San Antonio
Austin
Ft. Worth
Dallas

PLATEAU OF MEXICO

SIERRA MADRE ORIENTAL

San Luis Potosí
Aguascalientes
León
Santiago R.
Guadalajara
Lerma R.
Chapala
Morelia
Cuernavaca
Puerto Vallarta

Tampico
Veracruz
Orizaba (Citlaltépetl) 18,701 ft. (5,700m)
Popocatépetl 17,887 ft. (5,452m)
Mexico City
Puebla
Balsas R.
Oaxaca
Acapulco
ISTHMUS OF TEHUANTEPEC
SIERRA MADRE DEL SUR
Gulf of Tehuantepec

PACIFIC OCEAN *

800 miles
1,280 km
400

PACIFIC OCEAN

South America
Tropic of Cancer
Equator

# NORTH AMERICA: MEXICO & CENTRAL AMERICA I-

The seven Central American countries, south of Mexico, form a land bridge between North and South America. All but Belize and El Salvador have both Pacific and Atlantic coastlines. From mid-Mexico south, the climate is tropical: summerlike year-round except at higher elevations (where most of the population lives). Mountain ranges run the length of northern Mexico (the Sierra Madre Oriental in the east, the Sierra Madre Occidental in the west) and converge to form a high volcanic chain that continues through Central America. Violent earthquakes, hurricanes, and active volcanoes threaten the region. The fertile soil was enriched by past volcanic activity. Ample rainfall and sustained warmth produce fine mountain-grown coffee. Down along the steaming coast grow bananas, the second largest crop.

Except for the English influence in Belize, the language, culture, and religion of these nations reflect 300 years of Spanish domination, beginning in the early 1500s. Franciscan friars converted the Indians to Roman Catholicism and paved the way for their absorption into Spanish life. The indigenous populations were nearly wiped out by killings, enslavement, and particularly the European diseases to which they lacked immunity. "Mestizos," people of mixed Indian and Spanish heritage, make up most of the population. The few remaining pure-blooded Indian tribes live in isolated areas.

Most of these nations became independent from Spain in the early 19th century. But nearly 200 years later, the land, wealth, and political power are still in the hands of virtually the same wealthy families (the "landed aristocracy"). The vast majority of the people are desperately poor, and many try to migrate to the United States. The US has historically regarded this region as its sphere of influence, intervening in the affairs of these countries whenever it has felt a threat to its economic interests or political security.

## MEXICO A

**Area:** 761,602 sq. mi. (1,972,549 km²). **Population:** 86,500,000. **Capital:** Mexico City, 17,000,000. **Government:** Republic. **Language:** Spanish. **Religion:** Roman Catholic. **Exports:** Oil, vehicles, steel, chemicals, silver, coffee, cotton, sisal, and chicle. **Climate:** Temperate to tropical. ☐ Most of the world's largest Spanish-speaking population lives here, between the two Sierra Madre ranges, on a high triangular plateau. In the northwest, separated from mainland Mexico by the Gulf of California, is Baja (Lower) California, a long (800 mi., 1,280 km), narrow peninsula of mountains, deserts, and beaches. Mexico's large mestizo population (70%) is very proud of its Indian heritage. The 20th century Mexican muralists Rivera, Orozco, and Siqueiros have drawn heavily upon an ancient Indian muralist tradition. Long before the conquistadores destroyed the Aztecs, Mexico was home to the advanced Mayans and Toltecs, who built pyramids rivaling those in Egypt. Mexico City, the capital, is the world's largest (17,000,000), fastest growing, and smoggiest city. It was built on the site of the Aztec capital. Mexico is growing so rapidly that over half the population is under age 20. In the early 20th century, the nation had a revolution that set in place institutions for meaningful social, economic, and political reforms. Unfortunately, these measures have yet to be fully implemented as one political party has dominated Mexican politics since 1929. Mexico is mineral-rich: it is the leading producer of silver and may have the world's largest oil reserves. It is also the leading producer of sisal, a hemp fiber used in rope, and chicle, the basic ingredient of chewing gum. These substances grow in the rain forests of the low-lying Yucatán peninsula. Only 12% of Mexico is cultivated, but a wide variety of crops are grown, and produce is sold to US winter markets.

## BELIZE B

**Area:** 8,860 sq. mi. (22,947 km²). **Population:** 175,000. **Capital:** Belmopan, 4,600. **Government:** Constitutional monarchy. **Language:** English; some Spanish. **Religion:** Roman Catholic 60%; Protestant 40%. **Exports:** Sugar, timber, citrus, bananas, lobsters, and shrimp. **Climate:** Tropical. ☐ Located in the southeast corner of the Yucatán, Belize (beh leez'), formerly British Honduras but independent since 1981, is the only English-speaking country in this region. (Spain did not value the dense jungles and swampy coastlines.) Belmopan, the tiny capital, was placed inland for protection against hurricanes. Belize City (48,000) is an important port for tropical hardwoods (mahogany and rosewood). Half the people are black or mulatto (mixed black and white); a fifth are direct descendants of the Mayans; the remainder are mestizos, Europeans, and Asians. Though Guatemala has abandoned its claim to Belize, British troops still protect the nation's border.

## COSTA RICA C

**Area:** 19,620 sq. mi. (50,816 km²). **Population:** 2,950,000. **Capital:** San José, 248,000. **Government:** Republic. **Religion:** Roman Catholic. **Exports:** Coffee, bananas, timber, food products, and light industry. **Climate:** tropical. ☐ Among the countries in this region, Costa Rica has the highest standard of living; the highest percentage of mestizos (97%); the highest literacy rate; the greatest percentage of small landholders; and the longest orderly succession of democratic governments. It is the only Latin American country without a standing army—and this may have insured stability, since in Latin America the military is more likely to wage war against its own government than against any invader. Located on a coffee-growing plateau are the capital, San José, and other major cities.

## EL SALVADOR D

**Area:** 8,204 sq. mi. (21,249 km²). **Population:** 6,200,000. **Capital:** San Salvador, 465,000. **Government:** Republic. **Language:** Spanish. **Religion:** Roman Catholic. **Exports:** Coffee, cotton, sugar, timber, textiles, and food products. **Climate:** Tropical. ☐ This mountainous country is the smallest and most densely populated in the region and is the only one without an Atlantic coastline. El Salvador is more industrialized than its neighbors, but the creation of new jobs cannot keep pace with the expanding population. Most of the cities and farms are located in the central highlands region, where coffee is the principal cash crop. Over 90% of the people are mestizos, 3% are Indian, and 5% are the ruling white landowners. The enormous disparity between rich and poor has given rise to a revolutionary movement that controls many parts of the interior. The government, which has been supported by massive US aid, has been unable to defeat the rebels.

## GUATEMALA E

**Area:** 42,048 sq. mi. (108,904 km²). **Population:** 9,350,000. **Capital:** Guatemala City, 770,000. **Government:** Republic. **Language:** Spanish; many Indian dialects. **Religion:** Roman Catholic. **Exports:** Coffee, bananas, timber, cotton, chicle. **Climate:** Tropical. ☐ Guatemala has the largest percentage (55%) of pure-blooded Indians in the region. Most are direct descendants of the great Mayan culture, which lasted for nearly 2,000 years but ended mysteriously around 900 AD. Deep in the northern lowland jungles are the ruins of Tikal, a Mayan city of stone buildings and pyramids. The Indian majority live in their ancestral villages; the country is run by the westernized, mestizo Guatemalans ("ladinos"), who are concentrated in the southern highlands. Sitting on a high plateau is the capital, Guatemala City, the largest city in Central America. It has been wrecked by three devastating earthquakes in this century; the highlands are also prone to eruptions from some of the 27 volcanoes. For the past 30 years, a succession of military governments has waged a low-level war against guerrilla forces protesting the inequitable ownership of land. A civilian President was elected in 1986 but the war and the abuse of human rights continue. A third of Guatemalans have been converted to Protestantism by American evangelists. The Catholic Church in Latin America has been accused of being too sympathetic to the plight of the poor, and in countries with left-wing movements, the church has often been the target of the military. Evangelicals challenging Catholicism throughout Latin America have had the most success in Guatemala, which was formerly ruled by a military leader who became a "born-again Christian."

## HONDURAS F

**Area:** 43,270 sq. mi. (109,479 km²). **Population:** 5,000,000. **Capital:** Tegucigalpa, 310,000. **Government:** Republic. **Language:** Spanish. **Religion:** Roman Catholic. **Exports:** Bananas, coffee, timber, minerals, and cattle. **Climate:** Tropical. ☐ If any Latin American country deserves the name "banana republic," it is Honduras, the poorest country in the region. Huge, mostly American-owned plantations are located on the fertile and humid Caribbean coast. Here, the nation's only railroads are used for hauling bananas to coastal ports. Honduras, with large unplanted areas of cultivable land and large reserves of untapped mineral deposits, has significant economic potential. Tegucigalpa, the capital, is located in the mountains.

## NICARAGUA G

**Area:** 57,440 sq. mi. (148,770 km²). **Population:** 3,650,000. **Capital:** Managua, 680,000. **Government:** Republic. **Language:** Spanish. **Religion:** Roman Catholic. **Exports:** Coffee, cotton, coffee, sugar, bananas, and meat. **Climate:** Tropical. ☐ The largest country in Central America is triangular and made up of three distinct regions: the Mosquito Coast of swamps and rain forests; the mountains of the central highlands; and the fertile, hilly Pacific region, which holds Central America's largest lake, Lake Nicaragua. The lake is home to the world's only freshwater sharks, which evolved from sharks which were trapped when a volcanic eruption sealed off their bay from the ocean. The Pacific region holds the capital, Managua; the major cities; and the most productive farms. Managua was destroyed by earthquakes twice in this century. The population of Nicaragua is 85% mestizo, 10% mulatto, and 5% Indian. The Miskito Indians are mixed-blooded descendents of black slaves brought to Nicaragua during the early British rule of the Caribbean coast. In 1979, after 30 years of repression, Anastasio Somoza was overthrown by the left-wing Sandinista party. Fearing a communist foothold on the continent, the US organized a band of "Contras" to wage war against the Sandinistas. Though the rebels could not gain popular support, 10 years of war so weakened the economy that the Sandinistas yielded to US demands for free elections. In 1990, Violeta Chamorro led a coalition party to victory over the Sandinistas.

## PANAMA H

**Area:** 29,205 sq. mi. (75,641 km²). **Population:** 2,350,000. **Capital:** Panama City, 471,000. **Government:** Republic. **Language:** Spanish. **Religion:** Roman Catholic. **Exports:** Bananas, coffee, mahogany, and shrimp. **Climate:** Tropical. ☐ An average of 33 ships a day pass through the Panama Canal, earning Panama the title "crossroads of the world." Through this natural gap in the mountains, the Spanish used mules to pack Inca gold brought up from the west coast of South America. The United States helped create the nation of Panama as well as the canal. In 1903, Panama, with American support, asserted its independence from Colombia, which was opposed to the canal. The US-built canal, 50 mi. (80 km) long, was opened in 1914. In 1979, the US agreed to give Panama control of the Canal Zone, a strip 10 mi. (16 km) wide that crosses the isthmus. The US is to relinquish control of the canal in 1999. Most Panamanians live and work in the Canal Zone. Panama City is on the Pacific, Colón (78,200) on the Caribbean. In the late 1980s, the US began using economic sanctions to pressure Panama to depose its dictator, General Manuel Noriega. In 1989, the US invaded Panama on the grounds that the lives of American citizens and the operation of the canal were in danger. Noriega was taken to the US to stand trial for drug dealing.

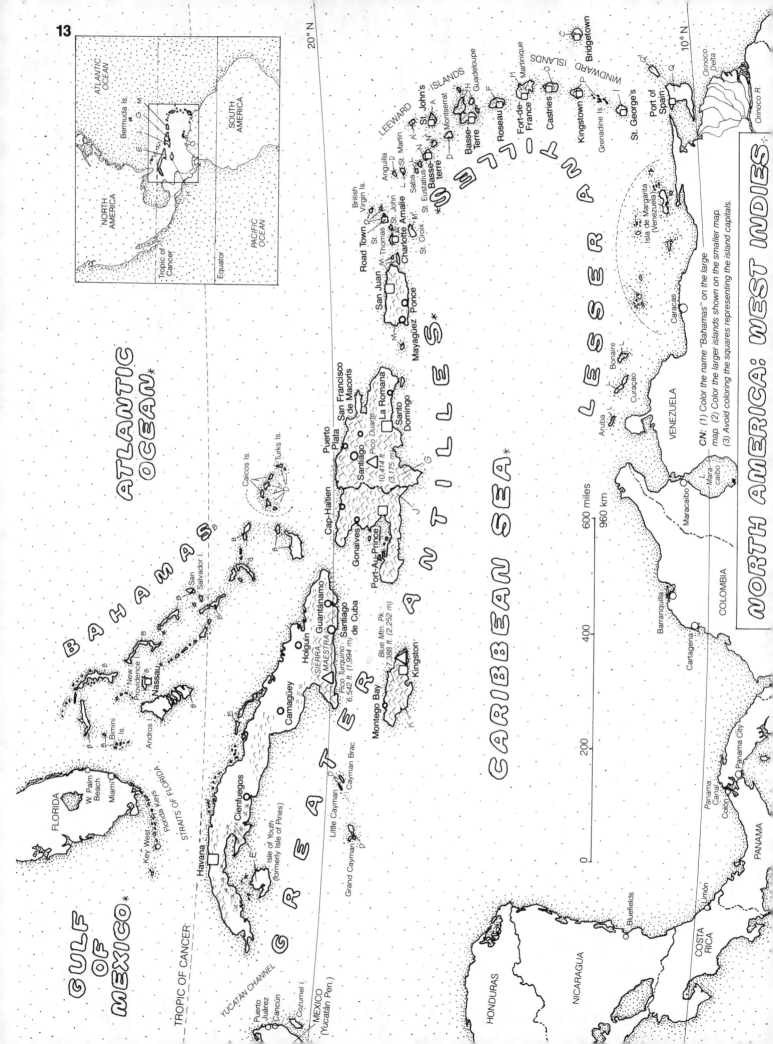

# NORTH AMERICA: WEST INDIES*

CN: (1) Color the name "Bahamas" on the large map. (2) Color the larger islands shown on the smaller map. (3) Avoid coloring the squares representing the island capitals.

Believing he had reached the East Indies, Christopher Columbus named these islands the "Indies." From 1492 to 1500 Columbus made four voyages to the West Indies, naming many of the islands and claiming them for Spain. Spain used these islands as a base for the exploration and plunder of the New World. A century later, other nations arrived to pirate Spain's treasure-laden galleons and contest the ownership of its colonies. Natives of the region were virtually wiped out by the Europeans. Those that did not succumb to Old World diseases died from overwork, beatings, and executions. Africans were brought to work the farms and plantations. Except in Cuba and Puerto Rico, the West Indies are mainly populated by pure or mixed-blooded (mulatto) descendants of black slaves. Catholicism dominates the former Spanish and French colonies; Protestantism is practiced on islands controlled by the British, Dutch, Danes, and Americans. Residents of Haiti practice voodoo. Many of these islands gained their independence after World War II.

The West Indies archipelago contains over 7,000 mostly uninhabited islands, cays (small islands), coral reefs, and rocks, which form the northern and eastern boundaries of the Caribbean Sea. The islands fall into three groups. (1) The Bahamas are coral islands off the coast of Florida. (2) The Greater Antilles, which include the larger islands of Cuba, Hispaniola (shared by Haiti and the Dominican Republic), Jamaica, and Puerto Rico, are the worn-down peaks of sunken mountains. (3) The Lesser Antilles are an arc of smaller volcanically formed islands. Trade winds, blowing from the east, keep the temperature of the tropical region comfortable. Agriculture flourishes in volcanic or alluvial soil. Sugar is king, followed by bananas. On most of these islands, tourism is the major industry.

## ANTIGUA & BARBUDA [A]

**Area:** 172 sq. mi. (445 km²). **Population:** 83,500. **Capital:** St.John's, 25,100. **Government:** Constitutional monarchy. **Language:** English. **Religion:** Protestant. **Exports:** Sugar, cotton, rum, and light industry. □ Antigua, the larger of the two islands, speaks English.

## BAHAMAS [B]

**Area:** 5,382 sq. mi. (13,939 km²). **Population:** 270,000. **Capital:** Nassau, 140,000. **Government:** Constitutional monarchy. **Language:** English. **Religion:** Protestant. **Exports:** Tropical fruit, fish products, and petrochemicals. □ Fifty miles (80 km) from Florida are the Bahamas, over 2,000 islands and coral reefs, only 29 of which are inhabited. Columbus may have landed in the new world on San Salvador Island. After the American Revolution, southern planters loyal to England fled to the Bahamas with their slaves. The country is still a haven, but now it's for foreign money.

## BARBADOS [C]

**Area:** 166 sq. mi. (430 km²). **Population:** 275,000. **Capital:** Bridgetown, 8,500. **Government:** Constitutional monarchy. **Language:** English. **Religion:** Mostly Protestant. **Exports:** Sugar, molasses, rum, and fish. □ Barbados, one of the world's most densely populated countries, is the most easterly of the islands. Residents speak with a British accent, play cricket, and drive on the left side of the road. Barbados has been called "England in the tropics."

## BRITISH TERRITORIES [D]

*Anguilla:* 35 sq. mi. (91 km²). Population, 7,250; capital, The Valley. *Bermuda:* 21 sq. mi. (53 km²). A group of islands north of the West Indies, about 850 mi. (1,360 km) north of Charleston, South Carolina. Population, 75,000; capital, Hamilton. *British Virgin Islands:* 59 sq. mi. (153 km²). Population, 12,500; capital, Road Town. *Cayman Islands:* 100 sq.mi.(259 km²). Population, 19,250; capital, Georgetown. *Montserrat:* 38 sq. mi. (98 km²). Population, 13,500; capital, Plymouth. *Turks & Caicos Islands:* 166 sq. mi. (430 km²). Population, 7,200; capital, Cockburn Town.

## CUBA [E]

**Area:** 44,219 sq. mi. (114,527 km²). **Population:** 10,400,000. **Capital:** Havana, 1,940,000. **Government:** One-party socialist republic. **Language:** Spanish. **Religion:** Roman Catholic. **Exports:** Sugar, tobacco, nickel, citrus, and fish. □ Only 90 miles (144 km) from Florida lies Cuba, the largest island in the West Indies. Cuba grows the world's largest export crop of sugar and its famous tobacco. After seeing natives smoking cigars, Columbus brought tobacco to Europe. The majority of Cubans are of Spanish descent. Fidel Castro's communist regime (the only one in the western hemisphere) has eradicated the poverty, hunger, disease, and illiteracy that characterize so much of Latin America. But it has resisted all attempts to democratize the system. Over 700,000 Cubans have migrated to the US since the revolution. A long-standing American trade embargo is a major reason for Cuba's dependence on Soviet aid.

## DOMINICA [F]

**Area:** 295 sq. mi. (764 km²). **Population:** 81,000. **Capital:** Roseau, 8,400. **Government:** Republic. **Language:** English. **Religion:** Mostly Roman Catholic. **Exports:** Bananas and coconuts. □ Dominica was named for Sunday (Domingo), the day that Columbus set foot on its soil. The extremely rugged terrain includes dense rain forests filled with exotic wildlife. Several hundred Caribs—the only remaining Indians native to the Caribbean—live here on a small reservation. The Caribs were a fierce tribe who migrated to the West Indies from South America; the word "cannibal" is derived from their name.

## DOMINICAN REPUBLIC [G]

**Area:** 18,750 sq. mi. (48,563 km²). **Population:** 6,750,000. **Capital:** Santo Domingo, 1,320,000. **Government:** Republic. **Language:** Spanish. **Religion:** Roman Catholic. **Exports:** Sugar, coffee, cocoa, food products, gold, and nickel. □ Christopher Columbus is believed to be buried in the capital, Santo Domingo, the oldest European city in the western hemisphere. Its university was established in 1535. The city was the headquarters for the early Spanish exploration of Latin America. The nation occupies two-thirds of the island of Hispaniola, which it shares with Haiti (it was part of Haiti until 1844).

## FRENCH TERRITORIES [H]

The French Antilles include: *Guadeloupe:* A group of 8 islands, 685 sq. mi. (1,774 km²). Population 335,000; capital, Basse-Terre. *Martinique:* 426 sq. mi. (1,103 km²). Population, 330,000; capital, Fort-de-France.

## GRENADA [I]

**Area:** 133 sq. mi. (344 km²). **Population:** 120,000. **Capital:** St. George's, 7,700. **Government:** Constitutional monarchy. **Language:** English. **Religion:** Roman Catholic; Protestant. **Exports:** Bananas, nutmeg, mace, and sugar. □ Grenada, the "Isle of Spice," made headlines in 1983 when the US invaded the island and overthrew the Marxist government, which it feared was an agent of communist expansion in Latin America.

## HAITI [J]

**Area:** 10,715 sq. mi. (27,752 km²). **Population:** 5,800,000. **Capital:** Port-au-Prince, 530,000. **Government:** Republic. **Language:** Haitian Creole; French. **Religion:** Voodoo; Roman Catholic. **Exports:** Coffee, sugar, and cocoa. □ Haiti, formerly the Spanish colony of Hispaniola, was seized by French pirates and later became a French possession. In 1804, plantation slaves revolted and Haiti became the Caribbean's first independent state and the world's first black republic. Corrupt and oppressive dictators (two of the more recent were father and son, "Papa Doc" and "Baby Doc" Duvalier) have made Haiti the poorest and most illiterate nation in the western hemisphere.

## JAMAICA [K]

**Area:** 4,244 sq. mi. (10,992 km²). **Population:** 2,480,000. **Capital:** Kingston, 650,000. **Government:** Constitutional monarchy. **Language:** English. **Religion:** Mostly Protestant. **Exports:** Bauxite, sugar, bananas, coffee, rum, and tobacco. □ Once the most important sugar and slave center in the region, Jamaica is now one of the world's leading producers of bauxite (aluminum ore). Heavy rainfall is responsible for the dense forests, rushing rivers, and cascading waterfalls. Jamaica's scenic beauty and carefree life style—which tourists find so attractive—mask the island's underlying poverty.

## NETHERLANDS ANTILLES [L]

The Dutch possessions consist of two groups of islands. (1) Close to Venezuela are *Aruba* (will be independent in 1996); 75 sq. mi. (193 km²). Population, 67,100; capital, Oranjestad. *Bonaire:* 111 sq. mi. (288 km²). Population, 9,700; capital, Kralendijk. *Curaçao:* 171 sq. mi. (444 km²). Population, 171,000; capital, Willemstad. Aruba and Curaçao have huge refining facilities for Venezuelan oil. (2) East of Puerto Rico are *Saba, St. Eustatius,* and the southern part of *St. Martin,* tiny islands with a total area of 26 sq. mi. (68 km²).

## PUERTO RICO [M]

**Area:** 3,435 sq. mi. (8,897 km²). **Population:** 3,750,000. **Capital:** San Juan. **Language:** Spanish; English. **Religion:** Roman Catholic. **Exports:** Pharmaceuticals, chemicals, light industry, sugar, bananas, and coffee. □ The US acquired Puerto Rico from Spain following the Spanish-American War of 1898. US aid and investments have made Puerto Rico the region's most industrialized island. Most Puerto Ricans are of Spanish descent. They are US citizens but cannot vote in presidential elections. Statehood is a controversial issue.

## ST. KITTS-NEVIS [N]

**Area:** 101 sq. mi. (262 km²). **Population:** 47,000. **Capital:** Basseterre, 15,000. **Government:** Constitutional monarchy. **Language:** English. **Religion:** Protestant. **Exports:** Sugar, cotton, and vegetables. □ St. Kitts-Nevis is a two-island nation lying east of Puerto Rico. The majority live on St. Kitts (also called St. Christopher). It was once the base for British operations in the Caribbean.

## ST. LUCIA [O]

**Area:** 237 sq. mi. (614 km²). **Population:** 132,000. **Capital:** Castries, 46,000. **Government:** Constitutional monarchy. **Language:** English. **Religion:** Roman Catholic. **Exports:** Bananas, coconuts, and arrowroot. □ The French patois which is also spoken is a reminder of the island's history; it alternated between British and French ownership before independence in 1978.

## ST. VINCENT & GRENADINES [P]

**Area:** 150 sq. mi. (389 km²). **Population:** 132,000. **Capital:** Kingston, 23,250. **Government:** Constitutional monarchy. **Language:** English. **Religion:** Roman Catholic. **Exports:** Bananas, coconuts, and arrowroot. □ St. Vincent and 100 islands of the Grenadine chain make up this nation.

## TRINIDAD & TOBAGO [Q]

**Area:** 1,990 sq. mi. (5,154 km²). **Population:** 1,340,000. **Capital:** Port of Spain, 64,000. **Government:** Republic. **Language:** English. **Religion:** Roman Catholic; Protestant. **Exports:** Oil products, asphalt, sugar, and rum. □ Trinidad has produced oil for over 100 years. It also refines oil from Venezuela, only 7 mi. (11 km) away. Pitch Lake is the world's largest deposit of natural asphalt. Calypso music, which began in Trinidad, is often played and sung to the beat of empty oil drums. Forty percent of the population are real Indians—descended from 19th century immigrants from India. Tobago has 5% of the population.

## VIRGIN ISLANDS (US) [M']

**Area:** 132 sq. mi. (342 km²). **Population:** 110,000. **Capital:** Charlotte Amalie, 12,500. **Government:** Self-governing territory. **Language:** English. **Religion:** Protestant. **Exports:** Refined petroleum products, light industry, and rum. □ The Virgin Islands were so named by Columbus because of their pristine beauty. In 1917, the US purchased these Virgin Islands which were owned by Denmark. St. Croix (kroy) is the largest and most industrialized island. St. John is mostly a national park. St. Thomas is the tourist's favorite.

# SOUTH AMERICA: THE COUNTRIES

## INDEPENDENT NATIONS
ARGENTINA<sub>A</sub> / BUENOS AIRES
BOLIVIA<sub>B</sub> / LA PAZ, SUCRE
BRAZIL<sub>C</sub> / BRASÍLIA
CHILE<sub>D</sub> / SANTIAGO
COLOMBIA<sub>E</sub> / BOGOTÁ
ECUADOR<sub>F</sub> / QUITO
GUYANA<sub>G</sub> / GEORGETOWN
PARAGUAY<sub>H</sub> / ASUNCIÓN
PERU<sub>I</sub> / LIMA
SURINAM<sub>J</sub> / PARAMARIBO
URUGUAY<sub>K</sub> / MONTEVIDEO
VENEZUELA<sub>L</sub> / CARACAS

## FOREIGN POSSESSIONS
FALKLAND ISLANDS<sub>M</sub> / STANLEY
FRENCH GUIANA<sub>N</sub> / CAYENNE

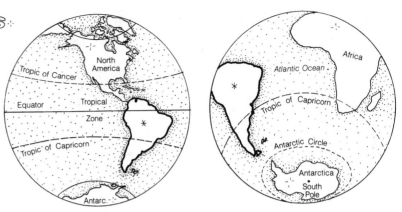

South America lies almost entirely east of North America, and most of it is within the tropics. Seasonal changes south of the equator occur at opposite times from those in the northern hemisphere (i.e., January is the warmest month). South America is much closer to Antarctica than any other continent.

South America, the fourth largest continent (6,900,000 sq. mi., 17,871,000 km²; population 300 million) is marked by stark contrasts: affluent cities are ringed by wretched slums; stone age cultures exist along with ultramodern cities; coastal areas are heavily populated while vast interiors remain uninhabited; rich and varied deposits of natural resources lie undeveloped because of inaccessibility; and there is an enormous disparity between the wealthy elite and the desperately poor.

Most of these nations are struggling with stagnant economies, overburdened by massive foreign debt obligations. This serious problem is compounded by a population explosion that is in some measure due to the influence of the Roman Catholic Church. The Church arrived shortly after the conquistadors and converted the native Indian population to the religion and culture of the European invaders. Natives were forced to work in the mines, farms, and settlements, and vast numbers died. Just as they destroyed the Aztec civilization in Mexico, a handful of heavily armed Spaniards conquered the flourishing 500-year-old Inca civilization. Descendants of the Incas are still the majority in Peru, Ecuador, and Bolivia. South America's earliest inhabitants are believed to have migrated from North America some 20,000 years ago. It may have taken another 10,000 years for them to reach the southern tip of the continent. Well over half of all South Americans are mestizos (of mixed Indian and European ancestry) who speak Spanish and are western oriented; mulattos (mixed black and European); and pure-blooded Indians and blacks. The remainder are of European descent. There were few Indians on the east coast; they were nearly wiped out by the colonialists, and Africans were imported as replacements. Settlers in Argentina and Uruguay encouraged European immigration instead of using slaves, and today the population in those countries is largely white. When slavery was abolished in the Guianas and blacks left the plantations, the British and Dutch looked to India and southeast Asia as a source for a new labor force. Guyana and Surinam are now dominated by descendants of those Asian indentured workers.

In the early 19th century, revolutionary fever swept the continent. European powers were too weakened by domestic wars to prevent their colonies from breaking away. New nations were formed under democratic constitutions, but elections have been rare. For nearly two centuries, extremely wealthy and influential families have been able to maintain enormous landholdings by encouraging dictatorial or military rule. The trend is now toward democratically elected governments, but in the absence of meaningful land reform, these emerging democracies are having to contend with political unrest bred by widespread poverty.

## 18TH CENTURY EUROPEAN COLONIES
SPAIN<sub>O</sub>
PORTUGAL<sub>P</sub>
GREAT BRITAIN<sub>Q</sub>
FRANCE<sub>R</sub>
NETHERLANDS<sub>S</sub>

In 1494, two years after Columbus discovered the New World, the Pope sought to avoid future conflict by drawing the "Line of Demarcation" down what was then believed to be the center of South America. He gave Portugal the rights to all lands east of the line, Spain, everything to the west. Subsequently, Portugal was allowed to expand its Brazilian colony westward. Neither nation coveted the swampy coast and rugged forests of the northeast, which were later claimed by the British (Guyana), Dutch (Surinam), and French (French Guiana), none of whom respected the Pope's territorial decrees (in South America or anywhere else). The shaded areas on the map below represent lands that Brazil later acquired from its neighbors.

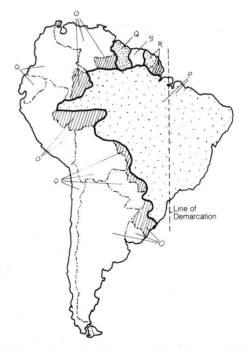

20° N

GREATER ANTILLES

LESSER ANTILLES

CARIBBEAN SEA *

ATLANTIC OCEAN *

Pico Bolívar
16,411 ft. (5,002 m)

Cristóbal Cólon
18,947 ft. (5,775 m)

Panama
Canal

10° N

L. Maracaibo

VENEZUELA

COLOMBIA

GUYANA

SURINAM

FR. GU.

Galapagos Is.
(Ecuador)

0°  EQUATOR

ECUADOR

Marajó I.

Chimborazo
20,561 ft. (6,267 m)

PERU

Huascarán
22,205 ft. (6,768 m)

10° S

PACIFIC OCEAN *

BRAZIL

L. Titicaca

Nevado Sajama
21,392 ft. (6,522 m)

BOLIVIA

1,500 miles

2,400 km

20° S

PARAGUAY

TROPIC OF CAPRICORN

Ojos del Salado
22,566 ft. (6,680 m)

1,000

30° S

Juan Fernández Is.
(Chile)

CHILE

Aconcagua
22,835 ft. (6,960 m)

ARGENTINA

URUGUAY

Río de la Plata

ATLANTIC OCEAN *

500

40° S

0

Falkland Is.
(Gt. Brit.)

Strait of
Magellan

Tierra del
Fuego

50° S

South Georgia I.
(Gt. Brit.)

Cape Horn

# SOUTH AMERICA: THE PHYSICAL LAND

*CN: (1) Use a dark or bright color for the major rivers (A, C, E, F, and G). Use lighter colors for their tributaries (B, D, H). (2) Color the highest peak in each west coast nation gray. (3) Color the major land regions on the small map below.*

## MAJOR RIVER SYSTEMS

AMAZON A
TRIBUTARIES B
ORINOCO C
TRIBUTARIES D
RÍO DE LA PLATA
PARANÁ E
PARAGUAY F
URUGUAY G
TRIBUTARIES H

## LAND REGIONS

ANDES MOUNTAINS I
GUIANA HIGHLANDS J
BRAZILIAN HIGHLANDS K
CENTRAL PLAINS
LLANOS L
SELVAS M
GRAN CHACO N
PAMPAS O
PATAGONIA P

Separating the highlands in the east from the Andes in the west are the central plains. The *llanos* are grassy cattle-grazing regions of Colombia and Venezuela. The *selvas,* the rain forests of the Amazon River Basin, cover parts of Brazil, Peru and Bolivia. The *Gran Chaco,* a generally dry and scrubby region of Bolivia, much of Paraguay, and part of Argentina, is known for the amazingly hard quebracho trees. The large, fertile, grassy, *pampas* region is Argentina's breadbasket. *Patagonia* is a cold, windswept area of deserts in southern Argentina.

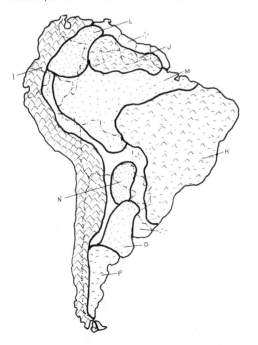

South America shares a common geology with Africa, from which it separated nearly 200 million years ago (Plate 2). Like North America, ice cream cone-shaped South America is broad in the north and narrow in the south. Both continents have three similar regions: ancient highlands in the east; younger and taller mountains in the west; and a wide central plain in between. In South America, the ancient highlands are the rugged Guiana Highlands in the northeast and the densely populated Brazilian Highlands in the eastern bulge. The Andes are the world's longest and second tallest mountain range (after the Himalayas). They extend 4,500 mi. (7,200 km) down the western edge of the continent, from the Caribbean to Cape Horn. The upper elevations have an eternal snow cover, even at the equator. Mt. Aconcagua (22,831 ft., 6,960 m) is the tallest of nearly 40 Andean peaks that are all higher than North America's Mt. McKinley (20,320 ft., 6,195 m). The Andes are still growing as the eastward-moving Nasca tectonic plate grinds under the South American plate (Plate 2). This is a region of geologic volatility, part of the "ring of fire" (Plate 2).

Dominating the central plains is the mighty Amazon River Basin of rivers, streams, and rain forests. The river is the world's largest and its length of 4,000 mi. (6,400 km) is second only to the Nile. From headwaters in the Peruvian Andes, the Amazon winds its way across the continent, transporting 20% of the world's total river water. The flow is so great that fresh water can be detected in the Atlantic 50 mi. (80 km) from the river's mouth. Parts of the river basin remain unexplored. Primitive tribes live there in complete isolation. The Amazon was named after the mythical Greek female warriors by a European explorer who claimed he saw women participating in a native battle. The rain forest (selva), which is the world's largest, is being destroyed at an alarming rate. This forest has been called the "earth's lungs," but trees actually perform the reverse function; they take in carbon dioxide and release oxygen. Scientists are concerned that the elimination of vast regions of oxygen-producing vegetation, along with the smoke from the burning brush, will add to the "greenhouse effect." The Amazon Basin has over 100,000 species of plants and animals; many are unique to the region and are in danger of extinction because of habitat destruction. South America is rich in distinctively different plant and animal life because of its isolation. The continent's largest native animal, the tapir, is no larger than a pony.

The two other important river systems are the Orinoco and the Río de la Plata. The Orinoco begins in the Guiana Highlands and makes a wide arc as it drains the grasslands of Colombia and Venezuela on its way to the Atlantic. The Río de la Plata system includes many rivers that provide essential transportation routes for Argentina, Bolivia, Brazil, Paraguay, and Uruguay. The Paraná, Paraguay, and Uruguay are the principal rivers of the system. They empty into the Atlantic through an *estuary* (an ocean inlet) called the Río de la Plata. Buenos Aires and Montevideo are located on opposite banks of this estuary. Brazil and Paraguay have completed the world's largest hydroelectric plant at the Itaipú Dam on the Paraná River. In the same region, separating Brazil and Argentina, are the Iguaçu Falls, the world's largest. Depending on the flow of the Iguaçu River, there are as many as 275 separate falls in this spectacular complex.

The continent's two major lakes are Maracaibo and Titicaca. Lake Titicaca, the largest, is in the Andes between Peru and Bolivia and at an altitude above 12,000 ft.(3,659 m) it is the world's highest navigable lake. The lake is also known for the basketlike sailboats that are woven from reeds growing along its shore. Even more remarkable are the floating islands woven from the same reeds. Houses are built on these islands.

On a narrow strip of land along the coast of Peru and northern Chile are some of the world's driest deserts. Parts of the 600 mi. (960 km) Atacama Desert in Chile have never recorded rainfall. This aridity, unusual on an ocean shore, is due to the interaction of the cold waters of the Humboldt (Peru) Current (Plate 44) and the Andes Mountains. Tropical ocean storms are cooled as they pass over the current, and they release their moisture before reaching the coast. The Andes block storms from the east. Further up the coastline (beyond the range of the Humboldt Current) is one of the world's wettest regions, the Pacific coast of Colombia. South America does not have the temperature extremes of North America, though it does have a steamy equatorial region, Andean glaciers, hot deserts in Argentina, and the cold southern tip of Tierra del Fuego. Temperatures on the southeast coast are moderated by the warm Brazil Current (Plate 44).

Cuba

GREATER ANTILLES

Jamaica

Haiti

Dom. Repub.

Puerto Rico

Virgin Is.

Anguilla

Barbuda

Antigua

St. Kitts-Nevis

Guadeloupe

Dominica

LESSER ANTILLES

Martinique

St. Lucia

Barbados

St. Vincent

Grenada

Tobago

Trinidad

CARIBBEAN SEA*

Texas

Idaho

Calif.

C

A

D

B

Wis.

Ind.

E

ATLANTIC OCEAN*

20° N

Cristóbal Cólon 18,947 ft. (5,775 m)

Santa Marta

Barranquilla

Cartagena

Panama Canal

Panama

Maracaibo

Barquisimeto

Valencia

Maracay

Caracas

L. Maracaibo

Pico Bolívar 16,411 ft. (5,002 m)

Arúca R.

Apure R.

Orinoco Delta

Ciudad Bolívar

Ciudad Guayana

Georgetown

New Amsterdam

Paramaribo

Kourou

Devil's Island

Cayenne

10° N

Minn.

Wis.

Mich.

Iowa

Ill.

Ind.

Bucaramanga

Medellín

ANDES MTS.

Cauca R.

Bogotá

Buenaventura

Cali

Magdalena R.

Meta R.

Orinoco R.

Angel Falls

Orinoco R.

GUIANA HIGHLANDS

C

D

B

ATLANTIC OCEAN*

Guaviare R.

Pico da Neblina 9,888 ft. (3,104 m)

AMAZON

Marajó I.

EQUATOR 0°

Quito

ECUADOR

Putumayo R.

Negro R.

Manaus

Amazon R.

Belém

São Luis

Iquitos

Amazon R.

Juruá R.

Amazon R.

BASIN

F

Fortaleza

PERU

Purus R.

Madeira R.

Tapajós R.

Xingu R.

Tocantins R.

Parnaíba R.

Cape São Roque

Natal

Recife

10° S

Lima

Araguaia R.

São Francisco R.

Maceió

Aracaju

Salvador

L. Titicaca

La Paz

BOLIVIA

BRAZILIAN

Brasília

HIGHLANDS

Goiânia

Belo Horizonte

PARAGUAY

Paraná R.

20° S

CHILE

Paraguay R.

Campinas

Itaipú Dam

Iguaçu Falls

São Paulo

Rio de Janeiro

TROPIC OF CAPRICORN*

1,000

Asunción

Iguaçu R.

Santos

ARGENTINA

Paraná R.

Curitiba

500

Santiago

Uruguay R.

Pôrto Alegre

Patos Lagoon

30° S

URUGUAY

Mirim Lagoon

F

Montevideo

Buenos Aires

Río de la Plata

PACIFIC OCEAN*

1,500 miles

2,400 km

ATLANTIC OCEAN*

F'

40° S

Brazil is nearly 10% larger than the United States, without Alaska and Hawaii.

# SOUTH AMERICA: NORTHERN COUNTRIES & BRAZIL

Though located in the heart of the tropics, the five northern countries of South America experience temperatures that vary from the heat of the steamy coastal lowlands to the cold of the snow-covered Andean peaks. In Colombia and Venezuela, most people live in the high valleys of three parallel extensions of the Andes. Residents of Guyana, Surinam, and French Guiana must live in the coastal heat because the interior of those nations is covered by the nearly impenetrable, and partly unexplored, Guiana Highlands. The presence of these highlands may have been the major reason why both Spain and Portugal ignored this region of South America. In the Venezuelan part of the Highlands is the world's tallest waterfall, Angel Falls (3,212 ft., 979 m), nearly 20 times higher than Niagara. It was unknown until 1935, when it was spotted from the air by an American pilot. The populations of Colombia and Venezuela have mixed Indian, Spanish, and black African roots. In both countries, 20% of the people are the white descendants of Spanish colonialists. As in other Latin American nations, this white minority owns the land and industries and controls the government. In the three smaller countries, fewer than 3% are white. Guyana and Surinam actually have Asian majorities, descendants of indentured workers from India and Indonesia who were brought to replace emancipated slaves.

## COLOMBIA A

**Area:** 439,750 sq. mi. (1,138,952 km²). **Population:** 32,000,000. **Capital:** Bogotá, 4,000,000. **Government:** Republic. **Language:** Spanish. **Religion:** Roman Catholic. **Exports:** Coffee, textiles, bananas, emeralds, and gold. **Climate:** Hot to freezing, depending on altitude. □ Colombia holds two geographical distinctions in South America: it is the only nation to have both Atlantic and Pacific coastlines, and it provides the only land entrance to the continent (from Panama). Colombia exports 90% of the world's supply of emeralds, but the nation's real treasure is its fine coffee beans—only Brazil grows more coffee. But in recent years, cocaine and marijuana have become the largest cash crop. Drug lords have resorted to widespread urban terror in resisting government attempts to curtail the drug trade. The focus of their activities is Medellín, one of three major cities located in the fertile Andean valleys. Bogotá is the capital and cultural center; Medellín is the finance and industrial center and the continent's leading textile producer. Colombia has the third largest population in South America. The eastern two-thirds of Colombia is a sparsely populated, grassy plain (llano) that merges with the Orinoco and Amazon Basins. Because land travel in the country is obstructed by natural barriers, Colombia was among the first nations to have a commercial airline.

## FRENCH GUIANA B

**Area:** 35,200 sq. mi. (91,168 km²). **Population:** 80,000. **Capital:** Cayenne, 39,000. **Government:** French Overseas Dept. **Language:** French. **Religion:** Roman Catholic. **Exports:** Bananas, shrimp, sugar, and hardwoods. **Climate:** Tropical. □ French Guiana (ghee ah' na) is the only foreign possession left on the continent. For nearly 100 years it was used as an overseas penal colony. Devil's Island, the most notorious of Guiana's three prisons, was a living hell for political prisoners sent from France. Another former prison colony, Kourou, is now the launching site for the European Space Agency's satellite program. Almost all the residents are black or Creole (of mixed black and French ancestry), and most live on the coast. The considerable mineral wealth in the extremely rugged Highlands is not readily accessible. This very poor possession relies upon France for continued support.

## GUYANA C

**Area:** 83,100 sq. mi. (215,229 km²). **Population:** 980,000. **Capital:** Georgetown, 190,000. **Government:** Republic. **Language:** English; Creolese; Hindi. **Religion:** Hindu 35%; Christianity 55%; Islam 10%. **Exports:** Bauxite, sugar, and rice. **Climate:** Tropical. □ Many things about Guyana (guy ah' na), formerly British Guiana, suggest an Asian country rather than one in Latin America. The Asian majority who control commerce have often been in conflict with the black minority (40%) who are descendants of slaves the Asians replaced. The blacks have held the political power and have run a socialist government. A language called Creolese, which borrows from all the ethnic groups, is spoken along with English. Many coastal areas, including Georgetown, the capital, are below sea level and are protected by a network of dikes and canals, many of which were built by the Dutch over 300 years ago. Georgetown is known for the white-painted wooden buildings common in many Caribbean island cities. Guyana is a world leader in the production of bauxite. In the 1970s, the nation attracted foreign religious cults. The Jonestown Massacre occurred when an American cult committed mass suicide at the urging of its leader.

## SURINAM D

**Area:** 63,240 sq. mi. (163,792 km²). **Population:** 405,000. **Capital:** Paramaribo, 72,000. **Government:** Republic. **Language:** Dutch. **Religion:** Christianity 35%; Hindu 30%; Islam 20%. **Exports:** Bauxite, aluminum, bananas, and timber. **Climate:** Tropical and damp. □ In the 17th century, the Dutch made what was surely the worst land swap in history: they gave New Netherlands and New Amsterdam (the state and city of New York) to the British in exchange for what is now Surinam. Formerly Dutch Guiana, Surinam (or Suriname) is truly the melting pot of South America. Among the many ethnic groups living in relative harmony are Hindus, Indonesians, mulattoes, blacks, native Indians, Chinese, and Europeans. Black descendants of slaves, called "bush negroes," live an African existence in the rugged interior. Like Guyana, Surinam is a leading producer of bauxite.

## VENEZUELA E

**Area:** 352,144 sq. mi. (912,052 km²). **Population:** 20,500,000. **Capital:** Caracas, 3,500,000. **Government:** Republic. **Language:** Spanish. **Religion:** Roman Catholic. **Exports:** Oil, iron and steel, coffee, sugar, cotton, and food products. **Climate:** Varies according to altitude. □ Venezuela means "little Venice" in Spanish. Early explorers were reminded of the Italian city when they first saw Indian villages perched on stilts in the shallow waters along the shores of Lake Maracaibo. The black ooze that sullies the lake eventually made Venezuela, in the 1930s, the world's first oil exporting nation and the wealthiest country in South America. But oil profits have enriched only the ruling elite. The government is saddled with a huge foreign debt, accumulated from loans made against future oil revenue that never materialized due to a drop in prices. Venezuela has the continent's largest iron deposits and an expanding steel industry. Caracas, the capital, is modern and prosperous, but like most other cities in Latin America it is surrounded by shantytowns. The government has enacted land reform and other programs to improve the quality of rural life and slow migration to the overcrowded cities. Much of Venezuela's land is drained by the Orinoco River. The river basin is an alluvial plain subject to both floods and drought.

## BRAZIL F

**Area:** 3,286,480 sq. mi. (8,511,983 km²). **Population:** 150,000,000. **Capital:** Brasília, 850,000. **Government:** Republic. **Language:** Portuguese. **Religion:** Roman Catholic. **Exports:** Coffee, bananas, soybeans, cotton, beef, timber, autos, machinery, and iron ore. **Climate:** Amazon region is tropical; northeast is subtropical; Brazilian Highlands region is temperate. □ The fifth largest nation in the world is the superpower of South America. Nearly half the size of the entire continent, and bordering all the other countries except Ecuador and Chile, it has over half of South America's population. Ten metropolitan areas each have over a million people. Brazil is the leading agricultural, mining, and industrial nation in Latin America. With less than 5% of its land under cultivation, Brazil still leads the world in production of coffee, bananas, and sugarcane. Most of the cane is converted into alcohol to fuel the nation's automobiles. About 30% of the world's coffee is grown in the southern Highlands, where the climate for this crop is ideal: hot, wet summers and mild, dry winters. The large coffee estates are called "fazendas," the Portuguese version of the Spanish "haciendas" that are so common in Latin America. Brazil ranks among the leaders in soybeans, beef, cotton, and timber. The nation is mineral-rich, having huge reserves of iron ore and deposits of nearly every important mineral. Despite this natural wealth, Brazil has become the Third World's largest debtor nation. Inflation is rampant, reaching 1,000% in the late 1980s. In 1990, the first democratically elected President in 30 years, Fernando Collor, instituted radical austerity measures. Brazil is a federal republic with 23 states, 3 remote territories, and a federal district, Brasília. In 1822, it gained independence from Portugal and became South America's only monarchy. It became a republic in 1889, but it has had a history of mostly totalitarian regimes.

With most of its population, wealth, and industry located on or near the Atlantic coast, Brazil can be compared to 19th-century America. But unlike Americans, Brazilians have been reluctant to move west, even though their government has constructed roads, offered free land, and even moved its capital to the futuristic city of Brasília, 600 mi. (965 km) from the coast. Major changes are taking place in the Amazon rain forest as enormous areas of vegetation are being stripped away to provide timber and grazing land. In response to international concern, Brazil has pledged to regulate the cutting down of the rain forest. Brazil is the only country in Latin America that has a Portuguese culture and language, but people of Portuguese descent make up only a small part of the population. Many Brazilians immigrated from other European, Muslim, and Asian nations. More Roman Catholics live in Brazil than in any other country in the world. About 40% of the people are nonwhite: caboclos (of mixed Indian and white descent); mulattoes (mixed black and white); and native Indians, some of whom live a stone age existence deep in the Amazon Basin. Except for the mistreatment of its native Indian population (a practice common to every developing nation), race relations in Brazil are generally good. The lack of education rather than race is usually the barrier to economic advancement.

The country is divided into 3 regions. (1) The Amazon River Basin is a warm and extremely humid rain forest, sparsely populated by Indian tribes. (2) The northeast is a forested plain that covers the Atlantic bulge; 30% of the population lives here. The plain is often ravaged by drought. (3) The central and southern plateaus (Brazilian Highlands) contain 75% of the agriculture, 80% of the mining and manufacturing, and over 50% of the population. This area has the best climate in Brazil. Located here are São Paulo (7,500,000), the fastest growing industrial center in Latin America, and Rio de Janeiro (5,200,000), a business and trade center. Rio is in a spectacular harbor setting, surrounded by a wall of mountains that nearly seals it off from the mainland. The city is best known for the most outrageous of the four-day carnivals celebrated throughout the nation before the Christian holiday of Lent.

With a huge landmass, a favorable climate, and untapped natural resources, Brazil has the potential to become a world superpower in the next century. Industrial production will be greatly aided by a network of hydroelectric plants going up around the country. Brazil and Paraguay have built Itaipú Dam, the world's largest power station, on the Paraná River. One hundred and twenty five miles to the south are the colossal Iguaçu Falls which offer even greater hydroelectric potential.

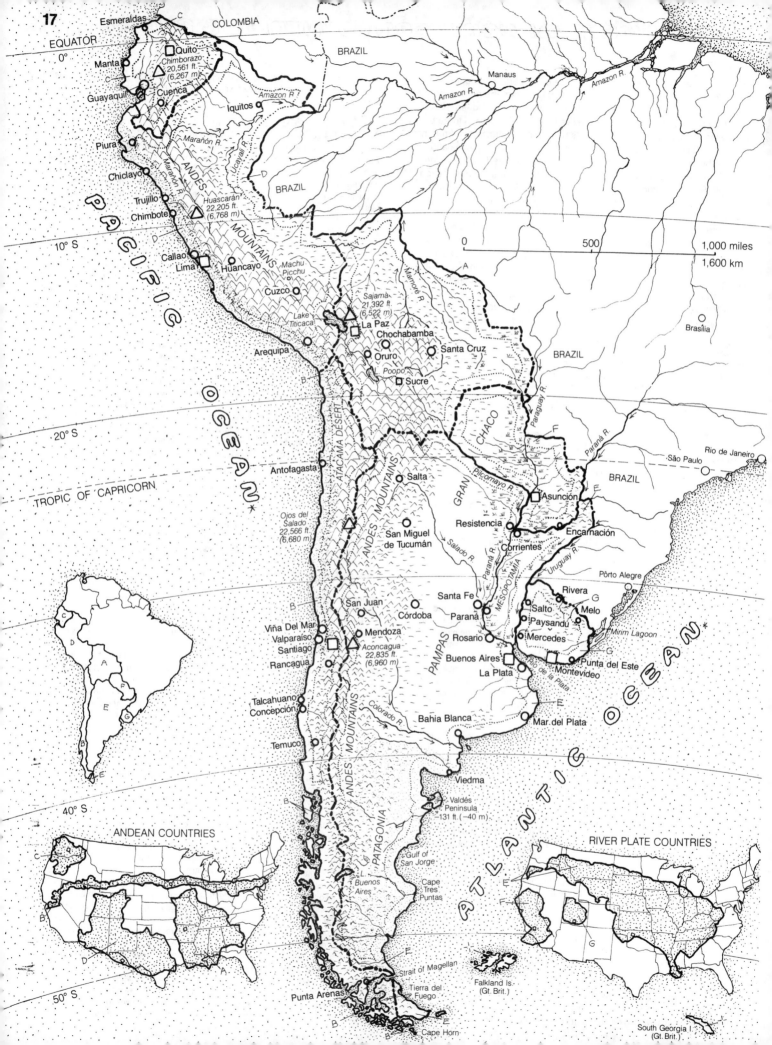

**17**

EQUATOR  0°

COLOMBIA

Esmeraldas
Manta
Quito
Chimborazo
20,561 ft.
(6,267 m)
Guayaquil
Cuenca
Piura
Iquitos
Marañón R.
Amazon R.

BRAZIL

Manaus
Amazon R.
Amazon R.

Chiclayo
Trujillo
Chimbote
ANDES
Marañón R.
Ucayali R.
Huascarán
22,205 ft.
(6,768 m)
BRAZIL

10° S
Callao
Lima
Huancayo
Machu Picchu
Cuzco
MOUNTAINS

0          500          1,000 miles
                        1,600 km

Arequipa
Lake Titicaca
Sajama
21,392 ft.
(6,522 m)
La Paz
Chochabamba
Oruro
Santa Cruz
Poopo
Sucre
Mamoré R.

Brasília

BRAZIL

PACIFIC
OCEAN

20° S

TROPIC OF CAPRICORN

Antofagasta
ATACAMA DESERT
Ojos del Salado
22,566 ft.
(6,680 m)
Salta
GRAN
CHACO
Paraguay R.
Pilcomayo R.
Asunción
Resistencia
San Miguel de Tucumán
Salado R.
Corrientes
Encarnación
Paraná R.
Paraná R.

Río de Janeiro
São Paulo

BRAZIL

Pôrto Alegre

San Juan
Córdoba
Santa Fe
Paraná
MESOPOTAMIA
Uruguay R.
Rivera
Salto
Melo
Paysandú
Mercedes
Mirim Lagoon

Viña Del Mar
Valparaíso
Santiago
Rancagua
Mendoza
Aconcagua
22,835 ft.
(6,960 m)
PAMPAS
Rosario
Buenos Aires
La Plata
Río de la Plata
Montevideo
Punta del Este

Talcahuano
Concepción
Colorado R.
Bahía Blanca
Mar. del Plata

ATLANTIC OCEAN

Temuco

ANDES MOUNTAINS

Viedma

Valdés Peninsula
−131 ft. (−40 m)

40° S

ANDEAN COUNTRIES

PATAGONIA

Gulf of San Jorge
Buenos Aires
Cape Tres Puntas

RIVER PLATE COUNTRIES

Strait of Magellan
Falkland Is.
(Gt. Brit.)

Punta Arenas
Tierra del Fuego

50° S

Cape Horn

South Georgia I.
(Gt. Brit.)

## ANDEAN COUNTRIES

The Andes divide Ecuador, Peru, and Bolivia into three regions: a western coastal desert, a central mountain range, and an eastern rain forest. In the Andes, people live at altitudes approaching 15,000 ft. (4,573 m). Life in the Andes is extremely hard. The mountainous terrain and the cold, oxygen-poor air have bred deep-chested people with short, stocky limbs. The staple of their meager diet is the potato, which may have originated here thousands of years ago. The llama, the "camel of the Andes," is a smaller, humpless relation of the camel; it provides transportation, wool, meat, and hides. The Andes nations have the largest Indian populations in the western hemisphere. Many are descendants of the highly evolved Inca civilization that flourished for 500 years before being destroyed by the Spanish.

### BOLIVIA A

**Area:** 424,100 sq. mi. (1,098,420 km²). **Population:** 7,000,000. **Capital:** La Paz, 890,000; Sucre, 79,500. **Government:** Republic. **Language:** Spanish; two Indian dialects. **Religion:** Roman Catholic. **Exports:** Natural gas, tin, coffee, silver, and cotton. **Climate:** Dry except in the northeast. □ With neither a coastline nor adequate road, rail, or water transportation, landlocked Bolivia is South America's poorest nation. On a plateau (altiplano) in the Andes sits the world's highest capital, La Paz (12,000 ft., 3,659 m), the seat of all government functions except sessions of the supreme court, which are held in Sucre, the official capital. The population is 50% Indian and 35% mestizo. Most Bolivians speak their own native languages. Only the wealthy and middle classes speak Spanish, the official language. The country was named after Simón Bolívar, the "George Washington of South America," a Venezuelan who helped bring an end to 300 years of Spanish rule. An unfortunate history of dictatorships and military coups was interrupted in 1952, when tinworkers led a revolution that brought in a reform-minded government. It was overthrown by the military in less than 12 years. Democracy returned in 1985. Bolivia lost much mineral-rich land, including a Pacific coastline, in border wars with Chile and Paraguay.

### CHILE B

**Area:** 292,250 sq. mi. (756,928 km²). **Population:** 13,000,000. **Capital:** Santiago, 4,300,000. **Government:** Republic. **Language:** Spanish. **Religion:** Roman Catholic. **Exports:** Copper, iron ore, chemicals, fruits, and fish products. **Climate:** Very dry to very wet. □ Picture a strip of land as long as the United States is wide but averaging less than 150 mi. (240 km) in width, and you have the dimensions of the world's narrowest nation, Chile (chee' lay). Rainfall varies from nothing at all in the mineral-rich Atacama Desert in the north to more than 200 in. (508 cm) per year on the stormy islands in the south. Chile is the world leader in copper production. Punta Arenas, close to Cape Horn, is the world's southernmost city. The eastern third of the country is occupied by the Andes. Most of the nation's activity is confined to the central valley region, a Mediterranean climate zone sandwiched between the coastal range and the Andes. Mestizos make up 70% of the population, but the 20% of European origin created the continental flavor of Chilean cities. Until a democratically elected Marxist was overthrown by the military in 1973, Chile had a tradition of political freedom, beginning with the liberal policies of the revolutionary leader and first president, Bernardo O'Higgins. During the recent period of military rule, ending in 1989, earlier gains in land reform were reversed by the return of confiscated property to the large landowners. Chile's Easter Island (Plate 33), famous for its huge stone figures of unknown origin, lies 2,300 mi. (3,680 km) to the west.

### ECUADOR C

**Area:** 108,000 sq. mi. (279,720 km²). **Population:** 10,500,000. **Capital:** Quito, 870,000. **Government:** Republic. **Language:** Spanish. **Religion:** Roman Catholic. **Exports:** Oil, bananas, coffee, and fish products. **Climate:** Varies according to altitude. □ Ecuador (ek' wa door) means "equator" in Spanish. Quito, the charming capital, lies only 15 mi. (24 km) from the equator, but at an altitude of 9,000 ft. (2,744 m), it enjoys being the "City of Eternal Spring." The hot, swampy coast produces the world's largest crop of bananas and feather-light balsa trees. Local palms supply the straw used in making Ecuador's famous "Panama hats." Almost all exports pass through the port of Guayaquil (1,225,000), the largest city and the commercial center. Politically conservative Quito is generally hostile to the progressive and liberal policies of Guayaquil. Mestizos make up 55% of the population, Indians 30%; most of the rest are the white elite. East of the Andes is the "Oriente," a primitive part of the Amazon River Basin. Oil from deposits in this area is piped over the Andes to the port of Esmeraldas. Ecuador owns the Galapagos Islands (Plate 14) 600 mi. (960 km) offshore, where plants and animals living in isolation have evolved into unique species. A visit by Charles Darwin in the 19th century helped formulate his theory of evolution.

### PERU D

**Area:** 496,200 sq. mi. (1,285,158 km²). **Population:** 22,000,000. **Capital:** Lima, 4,350,000. **Government:** Republic. **Language:** Spanish. **Religion:** Roman Catholic. **Exports:** Copper, lead, silver, zinc, fish products, coffee, and guano. **Climate:** Coast is mild and dry. East of the Andes is tropical. □ What was once the heart of the Inca civilization is now the third largest country in South America. The presence of Inca wealth induced the Spanish to found the coastal city of Lima (lee' muh) as the headquarters for their South American empire. The Andean city of Cuzco, the western hemisphere's oldest continuously inhabited city, was the Inca capital. Machu Picchu, at a higher elevation, was unknown to the Spanish. Its walled ruins were discovered

in the early 1900s. Over 9 million Inca descendants make up the largest Indian population of any country in the western hemisphere. Most live in poverty. Cities, farms, and factories are confined to coastal deserts, irrigated by runoff from the Andes. The cold ocean currents that prevent rain also support an abundant marine life. Peru is a leading exporter of fish, much of it in the form of animal feed made from ground anchovies. The fish attract seabirds, whose droppings (guano) contribute to a major fertilizer industry. Overfishing has led to a decline in the bird population, thereby reducing guano production. In the jungles of the northeast, at the headwaters of the Amazon, the port of Iquitos was built during the 19th century rubber boom. It still receives vessels that have sailed across an entire continent.

## RIVER PLATE COUNTRIES

The River Plate nations are united by South America's second largest river system. The Río de la Plata (River Plate) is an estuary of the Atlantic; the Paraná and Uruguay Rivers flow into it. The capitals of all three nations are located on this vital transportation system. Argentina and Uruguay, which have literate populations of European descent, are two of the continent's most prosperous nations. Paraguay, a landlocked nation with a mostly mestizo population, has struggled under years of military repression. Besides the Spanish language, one of the few things all three nations have in common is their love of maté, a locally grown herb tea. It is customarily served in a gourd and sipped through a straw (preferably made of silver).

### ARGENTINA E

**Area:** 1,075,500 sq. mi. (2,785,545 km²). **Population:** 32,500,000. **Capital:** Buenos Aires, 3,000,000. **Government:** Republic. **Language:** Spanish. **Religion:** Roman Catholic. **Exports:** Beef, hides, sheep, wool, grains, cotton, and food products. **Climate:** Mild in the north, cooler in the south. □ Argentina, the world's leading beef exporter, is the continent's second largest country in size and population. During the early part of the century, the sale of beef and cereals made it one of the world's richest nations. Under economic policies that began with Juan Perón in 1946, agriculture was heavily taxed to pay for social programs, and foreign investment was deterred by nationalization of industry. Such policies led to Argentina's current economic malaise. The population reflects its multinational origin. Spanish is said to be spoken with an Italian accent. Buenos Aires is a distinctly European city. There are five land regions in Argentina. (1) The western Andes contain the tallest peak in the western hemisphere, Aconcagua (22,835 ft., 6,950 m). (2) The scrub forests of the Gran Chaco lie in the north. (3) Mesopotamia, a damp agricultural region, is bordered by the Paraná and Uruguay rivers. (4) The treeless, grassy plains of the pampas contain fertile farmland, large ranches staffed by gauchos (cowboys), and major population and industrial centers. (5) Patagonia, a dry, inhospitable, windswept and thinly populated region, occupies the southern plateau. South of Patagonia is the island of Tierra del Fuego, which Argentina shares with Chile. Lying 300 miles (480 km) to the east are the Falkland Islands, a British possession which Argentina calls the Malvinas. In 1982, Argentina attempted to take the islands by force. Its defeat led to political turmoil at home and the overthrow of the military government.

### PARAGUAY F

**Area:** 157,047 sq. mi. (406,752 km²). **Population:** 4,100,000. **Capital:** Asunción, 460,000. **Government:** Republic. **Language:** Spanish, Guaraní. **Religion:** Roman Catholic. **Exports:** Beef, hides, tannin, coffee, and cotton. **Climate:** Warm and humid. □ The Paraguay (pa' ra gwy) River divides this landlocked country into two dissimilar regions: the Gran Chaco to the west and the Oriental in the east. The Paraguay and Paraná Rivers give the nation access to the Atlantic. Depending on the season, the Gran Chaco alternates between a dustbowl and swampland. In the Chaco grows the remarkable quebracho ("axbreaker") tree. Its virtually indestructible hardwood—too heavy to float—is used as railroad ties, telephone poles, and road-paving material. The tree also produces tannin (used in curing animal hides). Believing that the Gran Chaco contained oil, Paraguay seized it from Bolivia in a costly war in 1932. Most Paraguayans live in the gentle environment of the Oriental. They speak their native language, Guarani, as a second tongue. Paraguay has suffered under an endless succession of dictators. After the overthrow of General Alfredo Stroessner, who ruled for 35 years, an election was held in 1989.

### URUGUAY G

**Area:** 68,500 sq. mi. (177,415 km²). **Population:** 3,100,000. **Capital:** Montevideo, 1,300,000. **Government:** Republic. **Language:** Spanish. **Religion:** Roman Catholic. **Exports:** Meat products, hides, wool, and textiles. **Climate:** Mild and humid. Until the 1950s, Uruguay (oo' roo gwy) was a model society that was called the "Switzerland of South America." Its agricultural economy was prosperous, and its political and social policies were the most advanced on the continent. But when world commodity prices declined, Uruguay could no longer pay for its social bureaucracy. A new conservative direction led to political unrest and the formation of an urban guerrilla movement. The upheaval brought the military to power, and in the process of restoring order, Uruguay became a nation with the most political prisoners per capita. In 1985, a civilian government was elected. Uruguay's landscape is a gentle, grassy plain. The climate is always mild. A largely white population of Spanish and Italian extraction is the most urbanized (90%) in South America. Uruguay is one of the few nations in the world to have all its land in use.

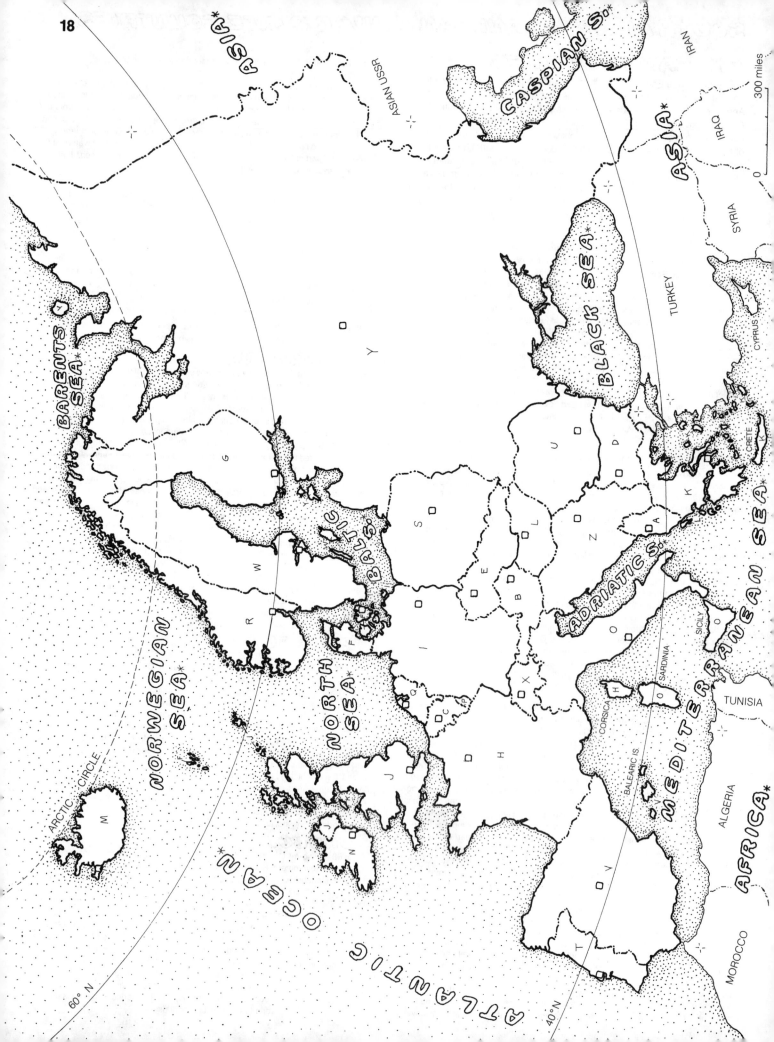

# EUROPE: THE COUNTRIES

CN: (1) You will probably have to use some colors more than once, so color a country first, then its name below. This will avoid having the same color on adjacent countries. (2) Color Europe gray on the two lower maps.

ALBANIA<sub>A</sub> / TIRANA

AUSTRIA<sub>B</sub> / VIENNA

BELGIUM<sub>C</sub> / BRUSSELS

BULGARIA<sub>D</sub> / SOFIA

CZECHOSLOVAKIA<sub>E</sub> / PRAGUE

DENMARK<sub>F</sub> / COPENHAGEN

FINLAND<sub>G</sub> / HELSINKI

FRANCE<sub>H</sub> / PARIS

GERMANY<sub>I</sub> / BERLIN

GREAT BRITAIN<sub>J</sub> / LONDON

GREECE<sub>K</sub> / ATHENS

HUNGARY<sub>L</sub> / BUDAPEST

ICELAND<sub>M</sub> / REYKJAVIK

IRELAND<sub>N</sub> / DUBLIN

ITALY<sub>O</sub> / ROME

LUXEMBOURG<sub>P</sub> / LUXEMBOURG

NETHERLANDS<sub>Q</sub> / AMSTERDAM

NORWAY<sub>R</sub> / OSLO

POLAND<sub>S</sub> / WARSAW

PORTUGAL<sub>T</sub> / LISBON

ROMANIA<sub>U</sub> / BUCHAREST

SPAIN<sub>V</sub> / MADRID

SWEDEN<sub>W</sub> / STOCKHOLM

SWITZERLAND<sub>X</sub> / BERN

USSR<sub>Y</sub> / MOSCOW

YUGOSLAVIA<sub>Z</sub> / BELGRADE

Europe is the mostly densely populated continent. Its 700 million people make up the second largest population (after Asia), but Europe is the second smallest continent (after Australia). Nine cities have over 2 million residents. Most of the population lives in the industrial regions of Great Britain, the Netherlands, Belgium, northern France, Germany, southern Poland, western Czechoslovakia, and northern Italy. The least populous region lies across northern Norway, Sweden, and Finland.

Only the largest 26 of Europe's 34 countries are included on this plate. (Liechtenstein, Monaco, San Marino, and Vatican City—shown on the following plates—are small enough to fit within the borders of any large city.) The European part of the USSR (Soviet Union) is nearly as large as the rest of Europe; the Asian part is three times as large. The USSR is considered European because most Soviets live in the European part and practice a European culture. Turkey is considered Asian because it has an eastern culture and only 3% of its area is in Europe.

Territorial barriers, in the form of mountains, rivers, lakes, gulfs, channels, and peninsulas, have enabled most European nations to preserve their individual cultures. Almost all Europeans speak some form of four Indo-European languages: (1) Celtic (Breton, Irish, Scottish Gaelic, and Welsh); (2) Latin-Romance (French, Italian, Portuguese, Romanian, and Spanish); (3) Germanic (Dutch, English, German, and the Scandinavian languages—Danish, Icelandic, Norwegian, and Swedish); (4) Slavic (Bulgarian, Czech, Polish, Russian, Serbo-Croatian, and Slovak).

Centuries of exploration, colonization, trade, and emigration have spread European culture across the globe. Beginning with ancient Greece, the birthplace of western civilization, Europe has been the source of great political ideas (democracy, capitalism, and communism); scientific and medical advances; technological achievements (the industrial revolution); and great art, music, and literature.

Christianity has been the principal religion, with Roman Catholicism the form most widely practiced (particularly in southwestern Europe). Protestant branches dominate Great Britain, Scandinavia, and northern Europe. The Eastern Orthodox branch of Christianity is the religion of the USSR and southeastern Europe. Judaism once flourished in Europe, but during World War II, Germany exterminated three-fourths of the Jewish population—6 million people.

The 40 years following World War II saw Europe divided into two armed camps: an Eastern Bloc of communist nations under Soviet domination and a western group of free, independent countries. Today, the continent is in a remarkable state of transition. Twelve of the western nations are members of the European Community, an organization that will tear down all trade barriers in 1992, creating one economic society. Future political unification (a "United States of Europe") has become a distinct possibility. In 1989, an even greater change occurred in Eastern Europe, when, with the acquiescence of the USSR, members of the Eastern Bloc asserted their independence by holding free elections and removing the communist party from power. The USSR, too, was opening up its society, adopting free-market incentives, and seeking economic ties with the West. The following year witnessed the reunification of the divided Germany and a general consensus that the "cold war" was indeed over.

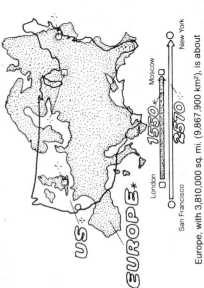

EUROPE*

US

San Francisco ○——□ London ◇——☆ Moscow
1550

2570
New York

Europe, with 3,810,000 sq. mi. (9,867,900 km²), is about 10% larger than the United States (Alaska and Hawaii included).

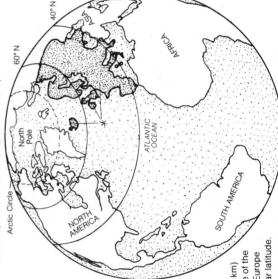

North Pole

Arctic Circle

60° N

40° N

ASIA

AFRICA

NORTH AMERICA

SOUTH AMERICA

ATLANTIC OCEAN

Europe is as far north as Canada, and London is 650 mi. (1,040 km) further north than New York City. But with the warming influence of the Gulf Stream (called the North Atlantic Drift in Europe) western Europe enjoys a milder climate than Canada and other regions of similar latitude.

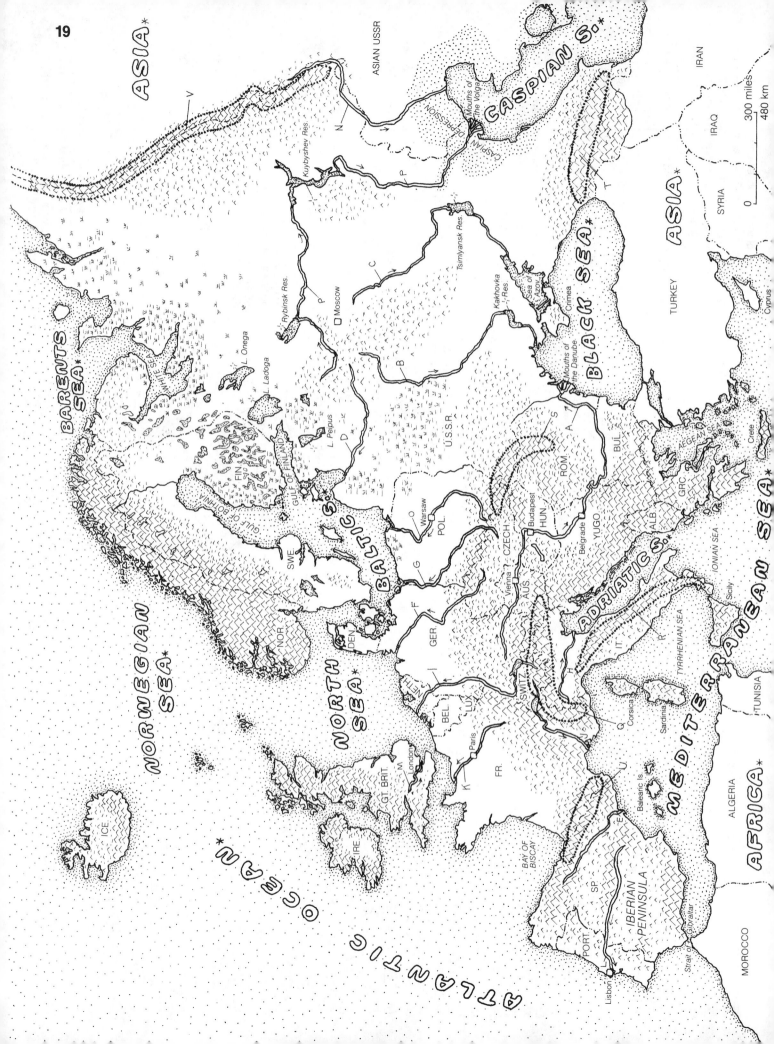

# EUROPE: THE PHYSICAL LAND -:-

CN: Use your lightest colors on land regions W–Z (on this page). (1) Color the rivers. (2) Color the principal mountain ranges (within the dotted outlines on the large map). (3) Color the principal land regions; then color gray the triangles representing famous mountain peaks.

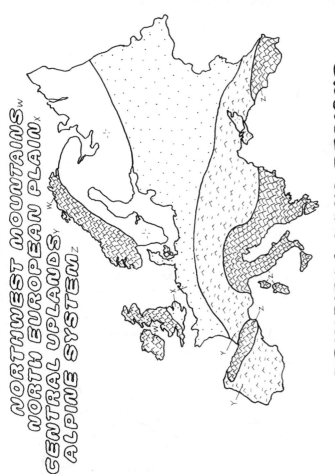

NORTHWEST MOUNTAINS w
NORTH EUROPEAN PLAIN x
CENTRAL UPLANDS y
ALPINE SYSTEM z

Europe is a collection of peninsulas that together form the western peninsula of the Asian landmass. Some geographers consider the two continents to be one: "Eurasia." The many peninsulas give Europe a longer coastline for its size than any other continent. With so many nations having access to the sea, Europeans have had a long history of shipbuilding, exploration, foreign trade, and fishing. The major fishing industries are on the Atlantic coast, not on the Mediterranean; the shallow Strait of Gibraltar bars the entry of the cold, deep Atlantic currents needed to sustain large fish populations.

Mountains play a major role in defining the landscape. The Urals in the USSR form Europe's eastern boundary. Mountains in the northwest cover most of Norway, Sweden, and Great Britain and part of Ireland. It is believed that these low mountains were part of the Appalachians when North America and Europe were joined, 200 million years ago (Plate 2). The larger, taller, and much younger Alpine System of ranges spans southern Europe from Spain to the USSR. The famous Alps contain over 1,000 glaciers and almost all of the continent's tallest peaks. But Europe's highest peak, Mt. Elbrus (18,480 ft., 5,634 m), is in the Caucasus Mountains, close to the Caspian Sea. The Caspian is actually the world's largest saltwater lake. Its surface is the lowest point in Europe (-90 ft., -27 m). The surrounding land, the Caspian Depression, is also below sea level.

## PRINCIPAL RIVERS -:-

DANUBE A
DNEPR B
DON C
DVINA (W) D
EBRO E
ELBE F
ODER G
PO H
RHINE I
RHÔNE J
SEINE K
TAGUS L
THAMES M
URAL N
VISTULA O
VOLGA P

## PRINCIPAL MOUNTAIN RANGES -:-

ALPS Q
APENNINES R
CARPATHIANS S
CAUCASUS T
PYRENEES U
URALS V

## PRINCIPAL LAND REGIONS -:-

Europe's most productive agricultural and industrial region is the Northern European Plain, which includes most of the European USSR, Poland, northern Germany, the Netherlands, Belgium, northern France, and southeastern Britain. This region has also been the site of Europe's bloodiest military battles; the flat lands and rolling hills are a natural avenue for invading armies. Also spanning the continent, lying between the Northern Plain and the Alpine System, is the less densely populated Central Uplands region of plateaus and rocky highlands. Most of its inhabitants live on small farms nestled in fertile valleys.

The continent's extensive river systems have historically provided important transportation routes. The longest river, the Volga (2,194 mi., 3,510 km), is the nucleus of a much larger river and canal network that services populated areas of the USSR and links the northern and southern coasts. The longest river in western Europe, the Danube (1,776 mi. 2,842 km), flows eastward through three capital cities. The Rhine (820 mi., 1,520 km), which flows northward through Switzerland, Germany, and the Netherlands, carries by far the most commercial traffic.

European climate varies from the damp, temperate western coast, moderated by the North Atlantic Drift (Gulf Stream), to the continental temperature extremes of the interior regions. The pleasant Mediterranean climate of southern Europe (mild, damp winters, and warm, dry summers) has given its name to similar climates in other parts of the world (California, central Chile, the cape of South Africa, and parts of Australia's south coast).

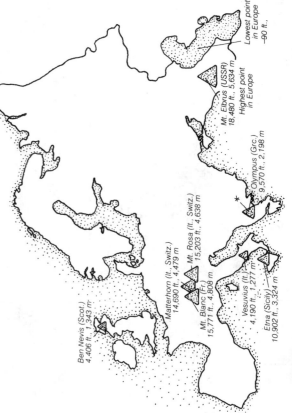

Ben Nevis (Scot.)
4,406 ft., 1,343 m

Matterhorn (It., Switz.)
14,690 ft., 4,479 m

Mt. Rosa (It., Switz.)
15,203 ft., 4,638 m

Mt. Elbrus (USSR)
18,480 ft., 5,634 m
Highest point
in Europe

Mt. Blanc (Fr.)
15,771 ft., 4,808 m

Vesuvius (It.)
4,190 ft., 1,277 m

Etna (Sicily)
10,902 ft., 3,324 m

Olympus (Grc.)
9,570 ft., 2,198 m

Lowest point
in Europe
-90 ft.

## PRINCIPAL MOUNTAIN PEAKS -:-

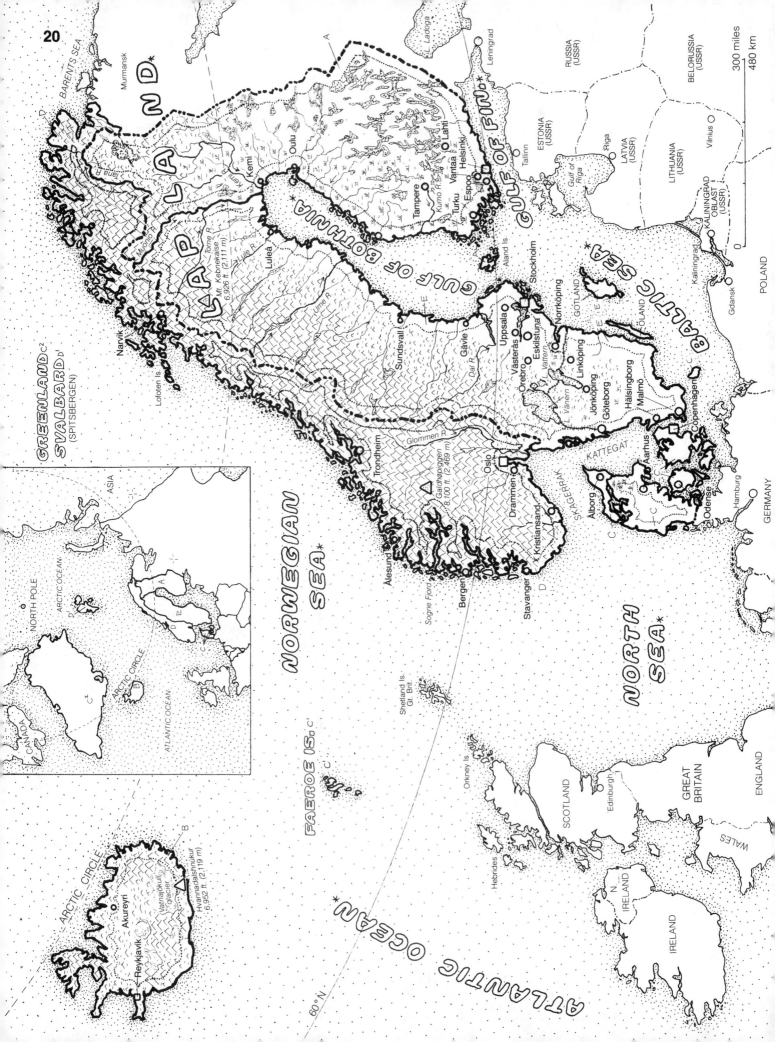

20

BARENTS SEA

Murmansk

GREENLAND C²
SVALBARD D¹
(SPITSBERGEN)

LAPLAND *

Tana R.
Muonio R.
Torne R.
Mt. Kebnekaise
6,926 ft. (2,111 m)

Narvik
Lofoten Is.

A

Oulu
Kemi

Lule R.
Lulea

Ume R.

Tampere
Turku
Vantaa
Helsinki
Espoo

GULF OF FIN *

Kumo R.
Lahti

Leningrad

RUSSIA
(USSR)

BELORUSSIA
(USSR)

300 miles
480 km

0

Tallinn

ESTONIA
(USSR)

Riga

Gulf of
Riga

LATVIA
(USSR)

LITHUANIA
(USSR)

Vilnius

KALININGRAD
OBLAST
(USSR)

GULF OF BOTHNIA

Sundsvall

Gävle

Dal R.

Uppsala
Västerås
Örebro
Eskilstuna

Stockholm
Åland Is.

Norrköping
GOTLAND

Vättern
Linköping
Vänern
Jönköping Göteborg
Hälsingborg
Malmö

ÖLAND

BALTIC SEA

POLAND

Gdansk

Kalmar

Trondheim

Glommen R.

Galdhøpiggen
8,100 ft. (2,469 m)

Oslo
Drammen
Kristiansand

Ålesund

Sogne Fjord

Bergen

Stavanger

SKAGERRAK
KATTEGAT

Ålborg
Aarhus
Odense

Copenhagen *
Hamburg

GERMANY

NORWEGIAN
SEA *

FAEROE IS. C¹
C¹

Shetland Is.
Gt. Brit.

Orkney Is.

NORTH
SEA *

Hebrides
SCOTLAND
Edinburgh

N.
IRELAND

WALES

IRELAND

GREAT
BRITAIN

ENGLAND

ATLANTIC OCEAN *

60° N

ICELAND

ARCTIC CIRCLE

Akureyri

Vatnajökull
Glacier

Hvannadalshnukur
6,952 ft. (2,119 m)

Reykjavik

NORTH POLE

ARCTIC OCEAN.

ARCTIC CIRCLE

ATLANTIC OCEAN.

CANADA

ASIA

California

# EUROPE: NORTHERN

*CN: (1) Use a bright color for Denmark (C) and the Faeroe Islands (C¹) on the large map. (2) In the northern region of the large map, color the word "Lapland" gray. (3) On the small map, note Svalbard (D¹) in the Arctic Ocean.*

Though only Denmark, Norway, and Sweden are Scandinavian, all five countries of Northern Europe share a common history and continue to maintain strong ties through membership in the Nordic Council. At one point, during the Middle Ages, they were all politically united under Danish rule. Finland's Asiatic language and ethnic origins are significantly different from those of its neighbors. The Scandinavian languages have Germanic roots and are similar enough to be understood in any of the three countries. The five nations are democratic, prosperous, socially progressive, highly literate welfare states with free-market economies. Curiously, Denmark, Norway, and Sweden have kept their monarchies. Scandinavia was the home of the Vikings, a fierce tribe of seafarers who explored the northeastern coast of North America (they preceded Columbus by 500 years) and raided the British Isles. From the 9th through the 11th centuries, the Danish and Norwegian Vikings explored the North Atlantic and North America while the Swedish Vikings went eastward to Russia.

The landscape of the northern nations shows strong evidence of ice age activity. Glacial action left thousands of lakes, rivers, fjords (fiords), islands, and glacial moraines (piles of debris). The ocean currents of the North Atlantic Drift keep the climate fairly temperate, except in the northern interiors. But almost every part of the region must endure long, dark winters. Scandinavians call the arctic zone (north of the arctic circle) the "Land of the Midnight Sun." The name refers to two summer months of continuous sunlight—winter brings with it two months of darkness. This northern region, also called Lapland, is home to some 60,000 Lapps who are short, colorfully dressed oriental-looking people. Lapps (or Samis, as they prefer to be called) have resisted all government attempts to assimilate them. Some still live the life of reindeer-herding nomads, roaming freely across international borders in the arctic region they call Sapmi.

## FINLAND A

**Area:** 130,120 sq. mi. (336,750 km²). **Population:** 5,000,000. **Capital:** Helsinki, 510,000. **Government:** Republic. **Language:** Finnish; Swedish (6%). **Religion:** Lutheran. **Exports:** Timber and paper products, plywood, manufacturing and engineering products, farmed furs, and textiles. **Climate:** Cold and snowy in the north. □ Finland is a land of 60,000 lakes and many more trees. In the southern lake and canal region, steamships service a 200 mi. (320 km) strip of countryside. Finland's prosperous economy emphasizes timber and paper products, including the world's largest plywood industry. Timbering is carefully monitored to protect the environment. Finland enjoys one of the world's highest rates of industrial growth. It exports its technology including entire factories. In the south, farms are surprisingly productive despite the short growing season. Finland excels in the design of fabrics, housewares, and furniture. It has produced two of the 20th century's finest architects: Alvar Aalto and Eliel Saarinen. The latter's son, Eero, became famous in America for such works as the Gateway Arch in St. Louis. Unlike the Scandinavians, the Finns are descendants of Asian settlers. Their language is related to Estonian and Hungarian, which have similar Siberian origins. The Finns are hardy outdoors people, passionately devoted to the ritual of the sauna bath: baking in a room heated by hot rocks, then bathing in icy water. Finland's intended capital was the site of the historic 1975 Helsinki Agreement, which intended to reduce East-West tensions in Europe, promote economic cooperation, and protect human rights. Throughout history, Finland has been ruled by Sweden or Russia. Finland has lost many wars and much territory to the Soviets, but since World War II it has been the only independent European nation on the Soviet border. The USSR is Finland's chief trading partner.

## ICELAND B

**Area:** 39,750 sq. mi. (102,953 km²). **Population:** 250,000. **Capital:** Reykjavík, 85,000. **Government:** Republic. **Language:** Icelandic. **Religion:** Lutheran. **Exports:** Fish products, aluminum, wool, and sheep products. **Climate:** Moderate. □ A former colony of Norway and Denmark, prosperous and socially progressive Iceland is Europe's westernmost nation (it is actually closer to North America). The treeless island of Iceland is not quite as cold as its name suggests though it does boast Europe's largest glacier, Vatnajökull. With the North Atlantic Drift passing by, harbors rarely freeze, and along some coastal areas a green ground cover is present most of the year. Iceland has been called the "Land of Ice and Fire." It is the most seismically active nation in the world and is constantly being enlarged as new islands are formed by volcanic eruptions (Plate 2). Homes are heated by hot springs. The word "geyser" comes from the name of one of Iceland's many hot springs, "Geysir," which intermittently spouts steam nearly 200 ft. (61 m) high. Fishing and the processing of fish products constitute the principal industry, and the island's territorial waters are zealously guarded. The literacy rate in Iceland (99.9%) is the world's highest. Reykjavík, the most northerly of all capital cities, has more bookstores per capita than any other city. The Icelandic language is remarkably similar to the Old Norse spoken by the Viking settlers. Iceland's parliament, the world's oldest form of representative government, was established by the Vikings in the 9th century.

## SCANDINAVIAN COUNTRIES

### DENMARK c

**Area:** 16,365 sq. mi. (42,352 km²). **Population:** 5,200,000. **Capital:** Copenhagen, 480,000. **Government:** Constitutional monarchy. **Language:** Danish. **Religion:** Lutheran. **Exports:** Meat, fish, dairy products, machinery, porcelain, pharmaceuticals, and furniture. **Climate:** Mild but damp. □ With few natural resources other than a low, flat, and fertile landscape enriched by glacial moraine, this tiny nation still provides its citizens a high standard of living. Danish foods, such as butter, cheese, bacon, and ham, are the profitable exports of a highly regulated agricultural industry. Cooperative farms are restricted in size, and farmers must pass licensing examinations. Denmark is a major exporter of housewares and furniture—"Danish modern" has become an international style. Most Danes live and work on the many ferry-connected islands east of the Jutland Peninsula, which Denmark shares with Germany. Except for the 40 mi. (640 km) border across the peninsula, Denmark is totally surrounded by water—it has almost 500 islands. On the easternmost island, ten mi. (16 km) from Sweden, is Copenhagen, the capital and cultural and industrial center. In its harbor is a statue of "the Little Mermaid," a character from a Hans Christian Andersen fairy tale. In the heart of Copenhagen is the famous Tivoli Gardens amusement park, the entertainment center of northern Europe. Denmark owns Greenland, the world's largest island, and the Faeroe Islands, which lie north of the British Isles. Both self-governing possessions were founded by Viking explorers.

### NORWAY D

**Area:** 125,051 sq. mi. (323,631 km²). **Population:** 5,200,000. **Capital:** Oslo, 470,000. **Government:** Constitutional monarchy. **Language:** two forms of Norwegian, plus many dialects. **Religion:** Lutheran. **Exports:** Oil, fish, timber products, chemicals, aluminum, and ships. **Climate:** Moderate along the coast and southern parts. □ Norway, the northernmost Scandinavian country, is a mountainous plateau with little farmland. Only 3% of Norway is cultivated; those farms are in the southern lowlands around Oslo, the capital. Oslo is the industrial, cultural, and recreational center. Norway's scenic western coastline has over 100,000 islands and a temperate climate. This coast is especially beautiful because of its majestic fjords (fiords)—long, narrow inlets of ocean that penetrate the steep coastal cliffs. The name "viking" comes from the Old Norse word "vik," which means "inlet." Sogne Fjord, the longest fjord, is 127 mi. (204 km) long. The 1,700 m. (2,720 km) coastline is actually many times longer if one includes the coasts of the thousands of islands and fjords. Fishing is no longer Norway's primary industry. It has been displaced by the drilling of oil and natural gas from the North Sea. Before the drop in energy prices in the 1980s, Norway was becoming quite wealthy. The nation also produces aluminum from imported bauxite. There is an incredible amount of energy available for manufacturing. In addition to oil and natural gas, Norway has the largest amount of hydroelectricity, per capita, in the world. The snowcapped mountains that supply the water-power also provide Norwegians with their favorite form of recreation: skiing.

### SWEDEN E

**Area:** 173,231 sq. mi. (443,124 km²). **Population:** 8,380,000. **Capital:** Stockholm, 675,000. **Government:** Constitutional monarchy. **Language:** Swedish. **Religion:** Lutheran. **Exports:** Machinery, timber, autos, and other transportation equipment. **Climate:** Moderate in the south. □ Both capitalists and socialists can admire Sweden, sometimes called the "Little of the Middle Way." Public and private ownership share in the profit-driven industrial economy, which provides its citizens with a high standard of living and a broad range of social benefits and services. The Swedes spend more on vacations, per capita, than any other nation. The majority of the people live in urban apartments, but many own second homes in the country. The long gray winters prevent Sweden from being the "perfect country." Sweden is the largest of the northern countries and occupies a long, narrow plain that slopes from the mountains it shares with Norway eastward to the Baltic Sea. Over half the nation is covered by forests, whose growth is monitored by a regulatory program as comprehensive as Finland's. Lake Vänern, the largest freshwater lake in western Europe, is the heart of a lake and canal linkage between Göteborg and the Baltic Sea just south of Stockholm. The lake is noted for the unusual and varied vegetation surrounding it—a reflection of the many kinds of soil created by ancient geologic activity. The northern coast, uninfluenced by the North Atlantic Drift, remains frozen 6 months of the year. Most Swedes live in the southern lowlands region. Located on the Baltic is Stockholm, the beautiful capital, which is also the commercial and cultural center. The city is built on 14 islands connected by bridges. Sweden's industry is based upon three abundant resources: iron ore, timber, and hydroelectric power. Unlike most industrial nations, Sweden manages to prevent its factories from violating the beauty of the landscape. Sweden, like Switzerland, maintains a large standing army, and like its neutral counterpart on the continent, it has avoided the great wars of this century.

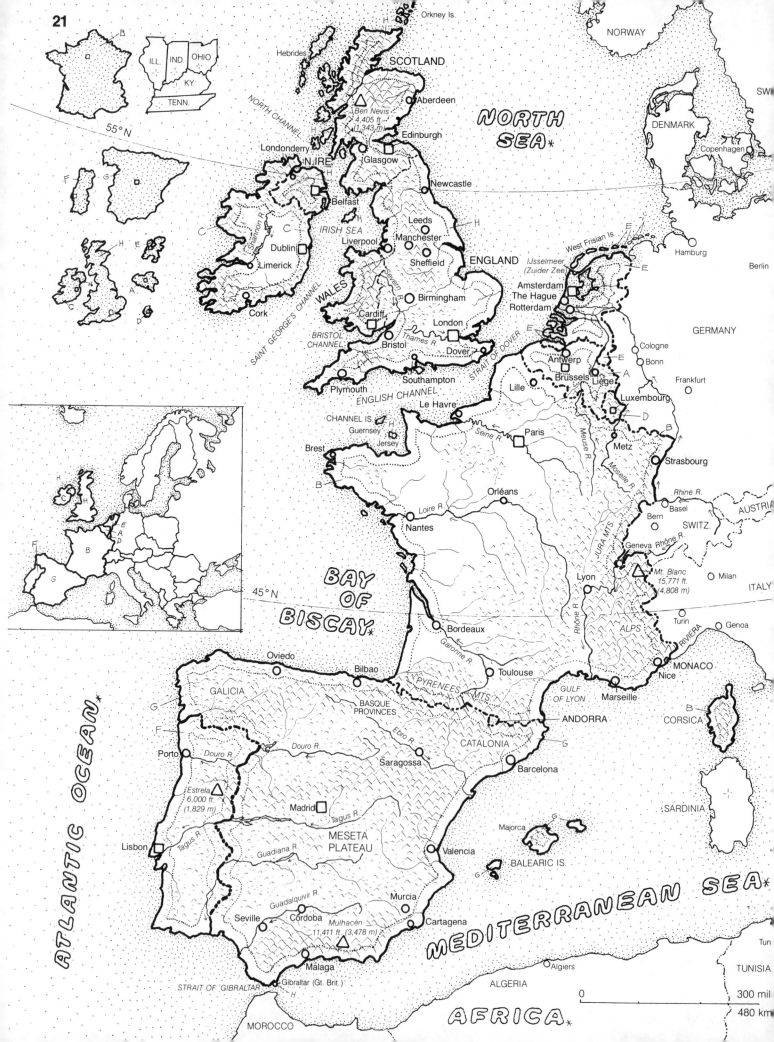

21

ILL. IND. OHIO
KY.
TENN.

55° N

NORWAY

Orkney Is.

SCOTLAND

Hebrides

NORTH CHANNEL

△ Ben Nevis
4,405 ft.
(1,343 m)

Aberdeen ○

Edinburgh ○

NORTH
SEA *

DENMARK

Copenhagen

Londonderry
N. IRE.
Glasgow ○ □

Belfast □

Newcastle ○

Hamburg ○

Berlin

Dublin □

IRISH SEA

Leeds ○
Manchester ○

Liverpool ○

Sheffield ○

ENGLAND

West Frisian Is.

GERMANY

Limerick ○

Shannon R.

Amsterdam
The Hague
Rotterdam

Cologne ○
Bonn ○

Cork ○

WALES

Severn R.

Birmingham ○

Cardiff □

London □

Antwerp ○

Brussels ○ Liège ○

Frankfurt ○

SAINT GEORGE'S CHANNEL

BRISTOL
CHANNEL

Bristol ○

Dover □

STRAIT OF DOVER

Luxembourg □

Plymouth ○

Southampton ○

Le Havre ○

ENGLISH CHANNEL

Lille ○

Metz ○

Strasbourg ○

Meuse R.

Moselle R.

CHANNEL IS.
Guernsey
Jersey

Seine R.

Paris □

Rhine R.

Brest ○

Orléans ○

Bern ○

Basel ○

AUSTRIA

SWITZ.

Nantes ○

Loire R.

BAY
OF
BISCAY *

JURA MTS.

Geneva ○ Rhône R.

△ Mt. Blanc
15,771 ft.
(4,808 m)

Milan ○

Lyon ○

Bordeaux ○

Garonne R.

Rhône R.

ALPS

Turin ○

Genoa ○

Oviedo ○

Bilbao ○

Toulouse ○

GALICIA

BASQUE
PROVINCES

PYRENEES

MTS.

GULF
OF LYON

RIVIERA

MONACO

Nice

Marseille ○

ANDORRA

CORSICA

ITALY

Porto ○

Douro R.

Douro R.

Ebro R.

CATALONIA

Barcelona ○

Saragossa ○

△ Estrela
6,000 ft.
(1,829 m)

Madrid □

Tagus R.

Majorca

SARDINIA

Lisbon □

Tagus R.

Guadiana R.

MESETA
PLATEAU

Valencia ○

BALEARIC IS.

ATLANTIC OCEAN *

Guadalquivir R.

Murcia ○

Seville ○

Córdoba ○

△ Mulhacén
11,411 ft. (3,478 m)

Cartagena ○

MEDITERRANEAN SEA *

Málaga ○

Gibraltar (Gt. Brit.)

STRAIT OF GIBRALTAR

ALGERIA

Algiers ○

TUNISIA

Tun.

MOROCCO

AFRICA *

0          300 mil
0          480 km

# EUROPE: WESTERN

*CN: (1) Color the British-owned Channel Islands off the coast of France. (2) Do not color Andorra, located between France and Spain. (3) Color the Mediterranean islands.*

## BELGIUM A

**Area:** 11,785 sq. mi. (30,499 km²). **Population:** 9,920,000. **Capital:** Brussels, 1,055,000. **Government:** Constitutional monarchy. **Language:** Flemish; French; German. **Religion:** Roman Catholic. **Exports:** Steel and engineering products, food, textiles, and glassware. **Climate:** Moderate; moist. □ Brussels, as the location for many economic, political, and military organizations, has been the de facto capital of western Europe since the end of World War II. It is the headquarters for the European Community, which is leading its member nations toward economic unification in 1992. Though steel is the largest of Belgium's many industries, the nation has the most productive (and smallest) farms in Europe. A source of continuing friction has been the division between the Flemings in the north, who speak a Dutch dialect, and the Walloons in the south, who speak French. Through the years, the flat, strategically located nation has been the site of numerous battles between invading armies. Napoleon "met his Waterloo" (his final defeat) near the Belgian village of that name.

## FRANCE B

**Area:** 212,467 sq. mi. (549,864 km²). **Population:** 54,820,000. **Capital:** Paris, 8,600,000. **Government:** Republic. **Language:** French; other dialects. **Religion:** Roman Catholic. **Exports:** Coal, iron ore, autos, food products, wine, and textiles. **Climate:** Moderate in the west, continental in the interior, and Mediterranean in the south. □ The largest European country west of the USSR is an unusual combination of sophistication and provinciality. The French have traditionally excelled in diplomacy, science, art, architecture, music, literature, fashion, wine, and cooking, but France is primarily a nation of farms, towns, and villages—it is the largest agricultural country in Europe. Picturesque regions are identified with famous wines (Champagne, Bordeaux, Burgundy, and so on). It is often said that there is France and there is Paris. The French capital is the premier tourist city, with its magnificent architecture, museums (including the Louvre), cathedrals (such as Notre Dame), parks, boulevards, restaurants, high fashion, and the 19th-century engineering marvel, the Eiffel Tower. France has become a European leader in high technology. It is the world's largest producer of nuclear power plants and transportation systems (TGV high speed trains), and a collaborator in commercial aviation design (Concorde and Airbus jets) and space exploration (Ariane rocket). The diverse French countryside includes coastlines that face three different seas (the North Sea, the Atlantic Ocean, and the Mediterranean Sea); farmland and plateaus in the interior; and the towering Alps (with Mt. Blanc) in the southeast.

**ANDORRA.** Located high in the Pyrenees with an area of 190 sq. mi. (465 km²) and a population of 40,000, this Catalan-speaking republic is governed by both France and Spain. This results in the duplication of such services as the mail, schools, and currency. Tourists enjoy the duty-free shopping and the fine ski resorts.

**MONACO.** With an area of 0.75 sq. mi. (1.9 km²), the word "tiny" fits this nation of 27,000 on the French Riviera. It is the home of many millionaires attracted by the elegant, tax-free living. Monaco made the news when actress Grace Kelly left Hollywood to marry Prince Rainier. His palace is in the capital, Monaco-ville. This nation also includes the city of Monte Carlo with its famed gambling casino.

## IRELAND C

**Area:** 27,120 sq. mi. (70,186km²). **Population:** 3,540,000. **Capital:** Dublin, 600,000. **Government:** Republic. **Language:** Gaelic and English. **Religion:** Roman Catholic. **Exports:** Meat and dairy products, textiles, technology, and whiskey. **Climate:** Mild and moist. □ The damp climate and limestone-rich soil have created the intensely green countryside that gives Ireland (Eire) the name "Emerald Isle." Tourism is the chief industry. Foreign investment in new factories halted 150 years of emigration (mostly to America), which began with the potato famine of the 1840s. Peat, a combustible soil, is the nation's primary fuel. It is dug out of damp, spongy bogs that cover a sixth of the country. The people of Ireland have a strong oral tradition, and though Gaelic is the traditional language, Ireland has produced some of the finest writers of English: Shaw, Swift, Joyce, Wilde, Yeats, and Beckett.

## LUXEMBOURG D

**Area:** 997 sq. mi. (2,580 km²). **Population:** 371,000. **Capital:** Luxembourg, 80,000. **Government:** Constitutional monarchy. **Language:** French; German; Letzeburgesch. **Religion:** Roman Catholic. **Exports:** Steel and chemical products. **Climate:** Continental. □ Higher than its "low" neighbors, Luxembourg, with its hills and valleys, forests, castles, and quaint villages, is also more scenic. Luxembourg, the capital city, is a center for banking and finance. A huge steel industry is at the heart of Luxembourg's prosperity. Citizens have to learn three languages: French for government affairs, German for writing, and Letzeburgesch, a German dialect, for conversation.

## NETHERLANDS E

**Area:** 16,040 sq. mi. (41,544 km²). **Population:** 14,710,000. **Capital:** Amsterdam, 800,000. **Government:** Constitutional monarchy. **Language:** Dutch. **Religion:** Roman Catholic 40%; Protestant 25%. **Exports:** Engineering and dairy products, natural gas, cut diamonds, and flower bulbs. **Climate:** Moderate and damp. □ Before Belgium and Luxembourg gained their independence in 1830, they were part of the Netherlands, a name which means "low countries." During the past 600 years, the Netherlands (Holland) has increased its size 40% by pumping out the North Sea. The reclaimed lands, called "polders," are protected by a network of dikes, ditches, and canals. Amsterdam, the capital, is built on a polder. Steam and electricity have replaced windpower as the energy source for the pumps. Windmills are still a familiar sight—along with fields of tulip bulbs—on the Dutch landscape. Bicycles fill the streets of Amsterdam, the center for trade, finance, and manufacturing. Rotterdam (560,000), on the Rhine River, is the world's busiest seaport.

## PORTUGAL F

**Area:** 35,500 sq. mi. (91,945 km²). **Population:** 10,200,000. **Capital:** Lisbon, 820,000. **Government:** Republic. **Language:** Portuguese. **Religion:** Roman Catholic. **Exports:** Cork, olive, and wood products; wines; and sardines. **Climate:** Mild; Mediterranean on south coast. □ The exports of western Europe's poorest nation are agricultural: cork, from the world's largest cork forests; olive oil; and fine wines, such as port and Madeira. Port is produced in the region surrounding Porto (335,000), the second largest city. Most people live in the wide, fertile coastal areas. Portugal's fine weather, church architecture, castles, and quaint villages draw an increasing tourist trade. Lisbon, on an inland sea formed by an estuary of the Tagus River, is one of Europe's most beautiful capitals. The Portuguese-owned Azores and Madeira Islands (Plate 36) are popular vacation resorts as well as agricultural regions.

## SPAIN G

**Area:** 194,884 sq. mi. (504,750 km²). **Population:** 39,410,000. **Capital:** Madrid, 3,800,000. **Government:** Constitutional monarchy. **Language:** Spanish; Basque; Catalan; Galician. **Religion:** Roman Catholic. **Exports:** Autos, wine, olive products, and cork. **Climate:** West is mild; interior is continental; southeast is Mediterranean. □ It isn't just good weather and low prices that make "sunny Spain" the most popular tourist attraction in Europe. Separated from the rest of the continent by the Pyrenees, and only 10 mi. (16 km) from North Africa, Spain has developed a unique blend of Western and Moorish cultures (the Moors occupied the country for over 700 years). Most tourists head for the Mediterranean coast, but many venture inland to see the bullfights, great art, medieval castles, and dramatic scenery. Spanish traditions, such as the siesta (a midday rest period) and the evening paseo (a walk before a late supper), are gradually giving way to the demands of a rapidly industrializing society. Spain shares the rugged Iberian peninsula with Portugal. Most of it is a craggy, high, dry, treeless plateau called the Meseta. All roads and rails lead to Madrid, the centrally located capital and the cultural, industrial, and commercial center of Spain. People in the distinctive regions of Galicia, the Basque provinces, and Catalonia speak their own languages (the ancient Basque language is related to no other) and have long sought independence from Spain. The exciting city of Barcelona (1,800,000), is Spain's second largest and the capital of Catalonia. The Spanish Empire was at its peak in the 16th century; its cultural influence is still evident in most of Latin America. At the mouth of the Mediterranean, the Rock of Gibraltar (1,398 ft., 426 m) occupies most of a narrow, 4 mi. (6.4 km) peninsula. Britain refuses to give up its naval base there and return Gibraltar to Spain.

## UNITED KINGDOM H

**Area:** 94,250 sq. mi. (243,919 km²). **Population:** 56,000,000. **Capital:** London, 7,300,000. **Government:** Constitutional monarchy. **Language:** English; some Welsh and Gaelic. **Religion:** Protestantism. **Exports:** Engineering products, autos, chemicals, food, and textiles. **Climate:** Temperate and moist. □ The British Isles include the United Kingdom (Great Britain and Northern Ireland) and the Republic of Ireland. The island of Great Britain includes the countries of England, Scotland, and Wales. The capital of the kingdom is London, England. The names "United Kingdom," "Great Britain," "Britain," and "England" are often used interchangeably. As recently as the early 20th century, the British Empire, the largest in history, governed a quarter of the earth. The empire over which "the sun never set" is nearly gone, but its influence remains. Britain still presides over the Commonwealth of Nations, an organization of 75 former colonies, current dependencies, and territories. The British Isles have been protected from continental interference by the English Channel, which was formed by rising seas following the last ice age, about 7,500 years ago. The last land invasion was the Norman Conquest of 1066. The rugged hills of Scotland in the north and Wales in the west are well suited for the grazing of livestock. Southeastern England has a significant amount of highly productive farmland. Wales has the richest coal reserves on the island. Its capital, Cardiff (279,000), and most of the population are located on a narrow coastal plain. Belfast (305,000) is the capital of Northern Ireland. This strife-torn region was called Ulster when it was the northern province of Ireland, before the southern part of the island won its independence from Britain in 1921. The continuing violence stems from Protestant rejection of demands made by the downtrodden Catholic minority, who want unification with Ireland. The proud, independent people who live in scenic Scotland share the Gaelic language and a common heritage with the Protestants who migrated to Northern Ireland hundreds of years ago. The stone buildings of Edinburgh (440,000) make the Scottish capital an exceptionally beautiful city. Glasgow (750,000) is Scotland's industrial center. Scotland has three sparsely populated, rugged archipelagos: The Hebrides, the Orkney Islands, and the Shetland Islands (Plate 20). South of England, and close to the French coast, are the Channel Islands. The largest, Jersey and Guernsey (homes of the famous breeds of cattle), are popular vacation spots as well as producers of specialty export crops. Densely populated England has 80% of the kingdom's people. The Midlands region is the country's industrial heartland, and Birmingham (1,050,000) is the largest industrial city. The industrial revolution, fueled by large coal and iron reserves, had its beginnings in 17th-century England. London, located on the historic Thames River, is the cultural heart of the United Kingdom and a center for world finance. The discovery of North Sea oil in the 1970s reversed a decline in the economy that began with the breakup of the empire after World War II.

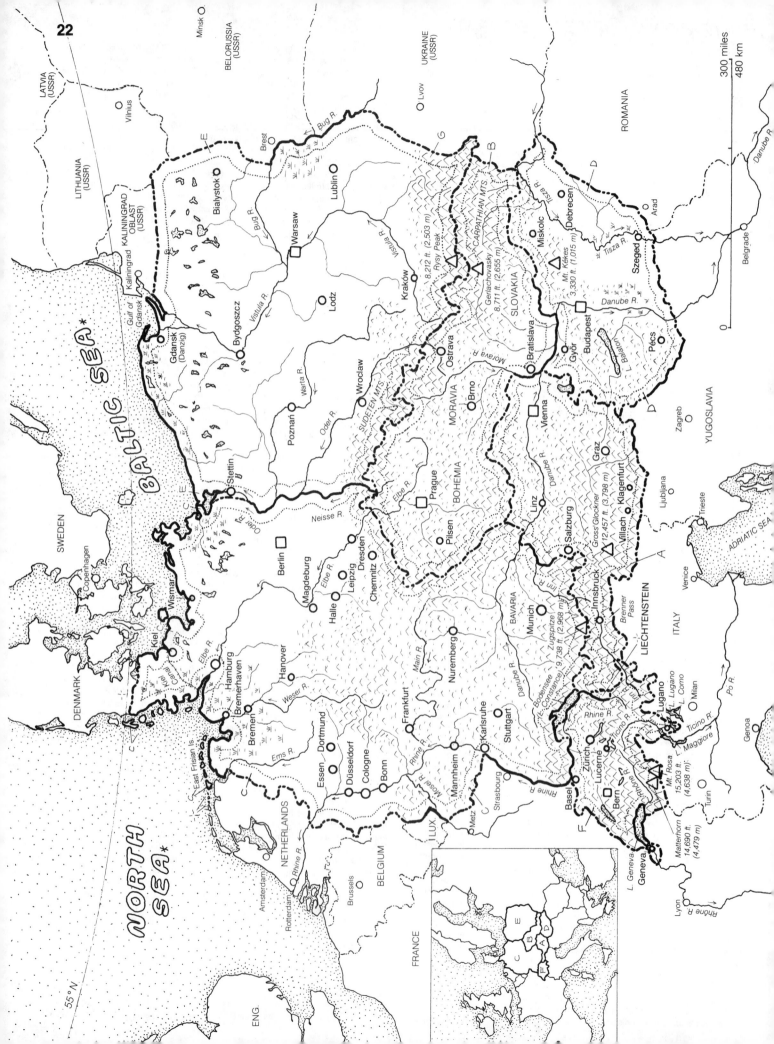

# EUROPE: NORTHCENTRAL

*CN: Do not color Liechtenstein, wedged between Switzerland and Austria.*

After its defeat in World War II, Germany was divided into two nations: East and West Germany. Europe itself was split into two armed camps: the free-market democracies of the West and the communist nations of the East (most of which were under Soviet domination). Austria, Liechtenstein, and Switzerland were in the western group, and Czechoslovakia, Hungary, and Poland were part of the Eastern Bloc. During the "cold war" period, the nations of the West prospered while the communist economies were barely able to provide necessities. In 1989, without opposition from the Soviet Union, a revolution occurred throughout the Eastern Bloc: the people rose up and threw their communist leaders out of power, paving the way for multiparty elections, the institution of free-market economies, and the restoration of religious freedom. Culminating these remarkable events was the reunification of Germany in 1990.

## AUSTRIA A

**Area:** 32,375 sq. mi. (83,850 km²). **Population:** 7,600,000. **Capital:** Vienna, 1,650,000. **Government:** Republic. **Language:** German (many dialects). **Religion:** Roman Catholic. **Exports:** Engineering, chemical, and forest products. **Climate:** Moist summers; cold, dry winters. ☐ Austria, even more than Switzerland, is dominated by the Alps; the towering peaks occupy 70% of the nation (versus 60% of Switzerland). Austria is famous for its ski resorts, music festivals, and health spas. Especially popular is the Mozart festival in Salzburg, the composer's birthplace. South of Innsbruck, on the Italian border, is the Brenner Pass, the main route across the Alps in central Europe and part of an amazing network of rail and road tunnels linking Austria, Switzerland, and Italy. Austria's central location gave it the title "Crossroads of Europe." Before World War II, Vienna, the home of Sigmund Freud, was the cultural, educational, scientific, and medical center of Europe. During the postwar period, it remained politically neutral, serving as a buffer between East and West. Many international organizations are headquartered in Austria. In the fertile Danube Valley, small, highly efficient farms supply most of the nation's needs. Austria's industry is also small and efficient and has the added advantage of cheap hydroelectric power generated by Alpine runoff. Austria is particularly concerned about preserving both its urban and rural environments: laws preventing destruction or defacement in town or countryside are strictly enforced.

## CZECHOSLOVAKIA B

**Area:** 49,373 sq. mi. (127,888 km²). **Population:** 15,800,000. **Capital:** Prague, 1,200,000. **Government:** Republic. **Language:** Czech 65%; Slovak 30%. **Religion:** Roman Catholic. **Exports:** Engineering, chemical, and wood products; textiles; and glass. **Climate:** Continental. ☐ Czechoslovakia came into existence in 1918, following World War I, when the two western Czech regions (Bohemia and Moravia) were combined with the eastern region (Slovakia). Each region has its own culture, and tensions have always existed between them. The nation's industrial output, among Eastern Bloc countries, was second to East Germany. Slovakia, with a third of the population, has traditionally been an agricultural region but is now industrializing. Rich coal deposits fuel the nation's heavy industry. Northern Bohemia is famous for fine glass products. An important crop is the special hop used in making Czechoslovakia's famed pilsner beer. Over 2,000 castles lie nestled in the thickly wooded countryside. During the summer, folk festivals are celebrated in quaint country villages. For many centuries, Prague, the beautiful capital, has been one of Europe's cosmopolitan centers. After the overthrow of the communist party, the Czechs elected the nation's leading playwright, Vaclav Havel, as their new president.

## GERMANY C

**Area:** 137,748 sq. mi. (356,866 km²). **Population:** 79,000,000. **Capital:** Berlin, 3,000,000. **Government:** Republic. **Language:** German and dialects. **Religion:** Roman Catholic in the south, Lutheran in the north. **Exports:** Autos, engineering and chemical products, lignite, foods, and textiles. **Climate:** Coast is moderate; inland is continental. ☐ No one can adequately explain how a nation that has excelled in literature, philosophy, science, medicine, and most notably, music (Bach, Brahms, Beethoven, Schubert, and Wagner) could have been capable of inflicting the horror that Germany caused in World War II. After its defeat, the victorious Allies divided Germany into two separate nations: East and West Germany. The prewar capital, Berlin, located in East Germany, was similarly divided. East Berlin was the capital of East Germany, and West Berlin was a detached part of West Germany within East Germany. With a massive infusion of Western aid, West Berlin became a showcase of capitalist achievement. As a result, East Berlin was forced to erect the infamous "Berlin Wall" to prevent its citizens from migrating to the West. This may have been the only wall in history designed to barricade its own people. The wall was torn down in 1990, after the fall of the communist regime and prior to the unification that ended over 40 years of division.

West Germany rose from the ashes of the war to become the world's fourth largest industrial economy. East Germany became the tenth largest, a more significant achievement because prior to the war the region was farmland, with only a quarter of Germany's population and few of the industrial centers and iron and coal deposits found in the western Ruhr and Rhine Valleys. The Ruhr is the largest industrial region in continental Europe. Germany's population is the largest in Europe outside the USSR. Most of the major cities are located in the west, on the bustling Rhine River. The southern Rhine Valley is known for its vineyards and castles. Hamburg (1,600,000), on the Elbe River close to North Sea, is the second largest city and busiest port; it is the cultural center of northern Germany. Munich (1,275,000), the capital of Bavaria in the south, though best known for beer festivals, has become a major industrial center. East Germany has been the world's leading producer of a low grade of soft coal called lignite. The Baltic coastal regions Germany shares with Poland are marshy and not suitable for farming.

## HUNGARY D

**Area:** 35,911 sq. mi. (93,032 km²). **Population:** 10,700,000. **Capital:** Budapest, 2,075,000. **Government:** Republic. **Language:** Hungarian. **Religion:** Roman Catholic. **Exports:** Transportation vehicles, pharmaceuticals, poultry, bauxite, and steel. **Climate:** Mild summers, continental winters. ☐ Hungarians, unlike their Slavic neighbors, are descendents of the Magyars, who came from central Russia in the ninth century. The Hungarian language, called Magyar (mod' jar), is related to Finnish and Estonian. The country is vertically bisected by the Danube River. To the east lies a low and flat agricultural plain—the site of an ancient sea. To the west is a hilly area which includes Lake Balaton, the largest lake in central Europe. It is surrounded by beaches and resorts. Two cities divided by the Danube, Buda and Pest, make up the nation's capital and industrial center. A quarter of the population lives in metropolitan Budapest. Hungary's feudal-based agricultural industry was improved by restructuring and the modern techniques introduced by the communists. But economic growth in the 1980s was due to Hungary's leadership among Eastern Bloc nations in pushing through the economic and political reforms that led to the revolution of 1989.

## POLAND E

**Area:** 120,732 sq. mi. (312,677 km²). **Population:** 37,000,000. **Capital:** Warsaw, 1,600,000. **Government:** Republic. **Language:** Polish. **Religion:** Roman Catholic. **Exports:** Engineering, steel, timber, and food products; coal; and sulphur. **Climate:** Continental away from the coast. ☐ Poland is a mostly flat country

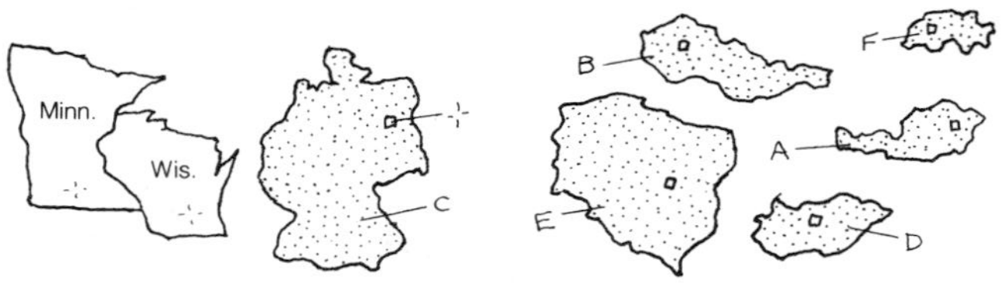


in the Northern European Plain. The land gradually rises from the Baltic coast until it reaches the mountainous border the nation shares with Czechoslovakia. Throughout its history, Poland has had its boundaries changed by invading armies. World War II began when Germany crossed Poland's western frontier. The Nazis were especially murderous in Poland: they killed 6 million Poles and 3 million Jews (most of Europe's largest Jewish community). Warsaw, the capital, was completely leveled. It is now one of the most modern cities in Europe, but the "old town" was faithfully restored. In the border adjustments after the war, Poland lost the eastern third of its land to the Soviet Union but gained the southern half of the prewar German province of East Prussia (a Baltic region detached from prewar Germany—the Soviets annexed the northern half, calling it Kalingrad Oblast); the free city of Danzig (now the port of Gdansk); and German lands east of the Oder-Neisse Rivers (the city of Wroclaw was formerly Breslau). In these border realignments, Poland was forced to give up to the USSR lands containing oil and potash, forests, and farms, but the country received from Germany an expanded Baltic coastline with two established ports, rich farmland, coal deposits, and a modern industrial region. Aided by enormous coal reserves, the economy changed from an agricultural to an industrial base. In the early 1980s, shipyard workers from Gdansk, led by Lech Walesa, formed Solidarity, an anticommunist union. They were aided by a strong Catholic Church. This led the way toward the election, in 1989, of another union member, Tadeusz Mazowiecki, as postwar Poland's first non-communist prime minister.

## SWITZERLAND F

**Area:** 15,941 sq. mi. (41,287 km²). **Population:** 6,500,000. **Capital:** Bern, 285,000. **Government:** Republic. **Language:** German 70%; French 20%; Italian 9%; many dialects. **Religion:** Protestant 52%; Roman Catholic 45%. **Exports:** Engineering and chemical products, scientific instruments, pharmaceuticals, watches, cheese, and chocolate. **Climate:** Temperature varies according to altitude. ☐ "Switzerland" is synonymous with "scenic," "skillful," "stable," and a standard of living that is the highest in Europe. The majestic Alps, which cover the southern half of the country, are a year-round tourist attraction. The pyramid-shaped Matterhorn (14,690 ft., 4,479 m) is one of the world's most striking peaks. The Alps are Europe's watershed: the Rhine flows to the North Sea, the Rhône to the Mediterranean, the Ticino (via the Po) to the Adriatic, and the Inn (via the Danube) to the Black Sea. Basel, on the Rhine, is the major port city. The nation's official language is German, but a majority speak a dialect called Swiss German. The nation's political stability (175 years of neutrality) has made Swiss banks a favorite depository for foreigners. Though the Swiss are ethnically diverse and live in almost completely independent cantons, they can become fiercely united on matters concerning economic prosperity and national security. Every adult Swiss male under age 50 is expected to belong to the militia. The Swiss obsession with neutrality is so strong, the nation refuses to join the United Nations for fear of having to take political positions.

**LIECHTENSTEIN.** South of Lake Constance, on the east bank of the Rhine and wedged between Switzerland and Austria, is a tiny (61 sq.mi., 158 km²) country of 23,000 German-speaking citizens. Liechtenstein has had close ties to Switzerland, which performs many services for its neighbor: the issuance of currency, mail delivery, telephone service, and so on. The nation's economy depends on tourists, light industry, the sale of postage stamps, and the residency of hundreds of corporations seeking a tax haven.

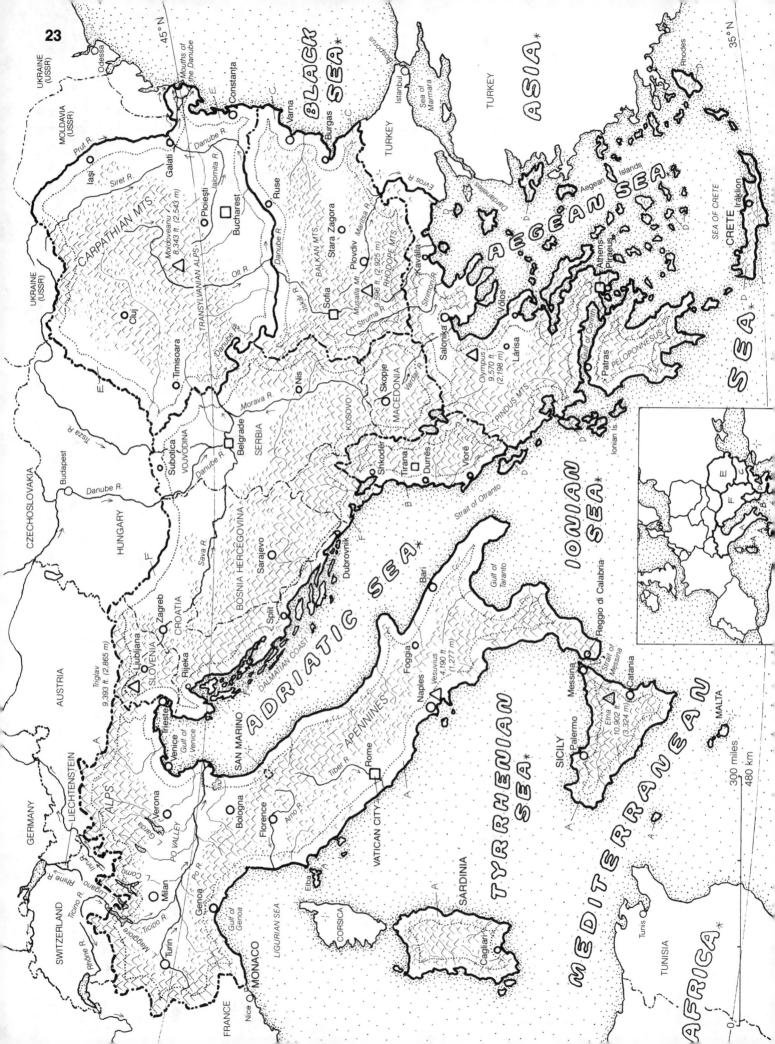

# 23

# EUROPE: SOUTHEASTERN

The coastal areas of southern Europe enjoy a Mediterranean climate of dry, hot summers and mild, moderately wet winters. These lands were once thickly forested, but centuries of logging, fires, overgrazing, and limited rainfall have stripped the landscape of almost all native trees. Though the land is only marginally fertile, with irrigation it can grow grapes, olives, figs, dates, citrus fruits, and chestnuts.

## ITALY

**Area:** 116,237 sq. mi. (301,054 km²). **Population:** 57,000,000. **Capital:** Rome. **Language:** Italian; many dialects. **Religion:** 2,850,000. **Government:** Republic. **Exports:** Engineering and food products, autos, plastics, wine, and silk. **Climate:** Mediterranean except in the north. Like Greece, Italy offers the visitor the antiquity of a great classical civilization (the Roman Empire) in a scenic setting. The landscape, climate, and waters around southern Italy are very much like Greece. In the north one finds additional attractions: great art, architecture, museums, opera, music festivals, and in the far north, alpine scenery. The Po Valley is the richest industrial and agricultural region in all of southern Europe—Italy has become the world's fifth largest industrial power. This wealthy northern region stands in stark contrast to the job-hungry south. Milan is the industrial and financial center; Turin is known for its canals, art, and architecture, and Florence is the center for Renaissance art. Close to the Swiss border are the beautiful lake regions of the Italian Alps. In the centrally located capital, Rome, the "Eternal City," are the historic ruins of the Roman Empire. Most of Rome's great art and architecture was commissioned by the Catholic Church, which is headquartered in Vatican City, a nation located within Rome. The only major city and industrial center in southern Italy is Naples. Located in particularly beautiful surroundings, it is 7 mi. (11.2 km) from the foot of Vesuvius. This still-active volcano is famous for an eruption that buried Pompeii and two other Roman cities. Only 2 mi. (3.2 km) from the toe of the Italian boot is Sicily, the largest island in the Mediterranean. Located there is Etna (10,902 ft., 3,324 m), Europe's tallest active volcano. Sicily is the home of the Mafia. Westward across the Tyrrhenian Sea is Sardinia, the Mediterranean's second largest island. Its culture bears the imprint of numerous invasions. On the mainland, close to the mainland, is the tiny island of Elba, the site of Napoleon's exile.

**VATICAN CITY.** Located within Rome is the smallest nation in the world. It is no larger than a few city blocks and has a population of 1,000. The Vatican holds St. Peter's, the world's largest Christian church, and the Vatican Palace, the residence of the Pope and the site of a glorious art collection that includes Michelangelo's Sistine Chapel ceiling.

**SAN MARINO.** Once a city-state in medieval Italy, San Marino, with a population of 23,000, is the world's oldest and smallest (24 sq. mi., 61 km²) republic. It lies nestled in the Apennines within sight of the Adriatic Sea. Like other postage stamp-sized countries, it derives revenue from selling stamps.

**MALTA.** About 370,000 people live on the three small islands (120 sq. mi., 310 km²) south of Sicily that form this nation. Through the centuries, Malta, the largest island, has been occupied by numerous invaders. Britain, the last nation to control Malta, used it as a naval base which is now a busy shipyard.

## BALKAN COUNTRIES

Albania, Bulgaria, Greece, Romania, and Yugoslavia occupy the Balkan Peninsula. The word "balkan" means "mountain" in Turkish. Agriculture is not very productive because of the rugged environment and outmoded methods of farming. In many rural villages, donkeys are still the principal form of transportation. Since World War II, Greece has been the only non-communist country in the region, but

in 1989, the revolutionary fervor of eastern Europe reached the Balkans and communism was under attack in every state but Albania. Regional elections held in 1990 did not completely remove communists from power as they did in eastern Europe. The depressed economies have only reinforced the Balkan Peninsula's reputation for being "Europe's poorest corner." The Balkans have also been an area of intense ethnic and religious hatred, both within and between nations. These attitudes were exacerbated during the post-World War I upheaval, when national boundaries were significantly altered and minority populations were trapped within hostile countries. Years of communist control suppressed but did not extinguish the antagonisms that have surfaced with the weakening of communist authority.

## ALBANIA

**Area:** 11,100 sq. mi. (28,748 km²). **Population:** 2,875,000. **Capital:** Tirana, 171,000. **Government:** Republic (communist). **Language:** Albanian. **Religion:** Moslem 70%. **Exports:** Food products, wine, and minerals. **Climate:** Mediterranean along the coast; continental inland. □ Albania is the smallest Balkan country in size, population, and per capita income. For over 40 years, this small mountainous nation has had the most repressive of all communist governments—one that severed relations with other communist nations because of their "liberal" policies, thus isolating itself from both the democratic and the communist worlds. In 1967, all mosques and churches were closed and the country became the world's first atheistic state. Until then Albania had been the only European nation with a Muslim majority—the result of 500 years of Turkish rule. In the late 1980s, Albania, under a new leader, began to soften its hard-line policies.

## BULGARIA

**Area:** 42,823 sq. mi. (110,912 km²). **Population:** 8,940,000. **Capital:** Sofia, 1,080,000. **Government:** Republic. **Language:** Bulgarian. **Religion:** Eastern Orthodox. **Exports:** Food products, grains, metals, textiles, and high technology. **Climate:** From continental to Mediterranean. □ Even as it shifts to an industrial economy, Bulgaria continues to be the Balkan Peninsula's leading agricultural nation and eastern Europe's winter breadbasket. Bulgaria is the world's largest supplier of attar of roses, an oil used in the manufacture of perfume. Byzantine-style architecture, cultural artifacts, and the consumption of yogurt (solidified, fermented milk) are reminders of 500 years of Turkish rule. But unlike the case in Albania, Islam did not win lasting converts. The Russians drove the Turks out in 1878, and now the majority of Bulgarians are Christians, followers of the Eastern Orthodox Church. Memory of the hated Turkish rule is so deeply embedded in the national consciousness that in 1984 the government forced the Turkish and Muslim minorities—1.5 million people—to Bulgarize their names and refrain from speaking Turkish in public. This policy was reversed in 1990.

## GREECE

**Area:** 50,943 sq. mi. (131,840 km²). **Population:** 10,000,000. **Capital:** Athens, 2,750,000. **Government:** Republic. **Language:** Greek. **Religion:** Eastern Orthodox. **Exports:** Food products, olives, wine, and citrus fruits. **Climate:** Mediterranean. □ Over 400 sun-drenched islands, deep blue skies, an unrivaled historic past, good food, music, and dance all are part of the Greek attraction. Almost all the islands in the Aegean archipelago, including most of those off the Turkish coast, belong to Greece. Crete, the largest island, was the site of the highly advanced Minoan civilization, which flourished 5,000 years ago. Most of Greece is a ragged peninsula whose southern tip, the Peloponnesus, is detached from the mainland by the Gulf of Corinth and a canal. The Peloponnesus was the site of the Persian Wars and the battles between Athens and Sparta. The Greeks have the oldest maritime tradition in Europe. Today, they operate the supertankers of the world's largest merchant

fleet. The Acropolis ("high city"), the heart of ancient Greece, is located on a hill in the middle of Athens. These ruins of structures built 2,500 years ago represent the birthplace of western culture and democratic ideals. Though modern Greece is rapidly becoming industrialized, agriculture remains a significant part of the economy. Wheat, cotton, and tobacco are grown in addition to the usual Mediterranean crops.

## ROMANIA

**Area:** 91,699 sq. mi. (237,500 km²). **Population** 22,750,000. **Capital:** Bucharest, 1,850,000. **Government:** Republic. **Language:** Romanian. **Religion:** Eastern Orthodox. **Exports:** Engineering, chemical, and food products; oil. **Climate:** Continental. □ Romanians trace their ancestry and their Latin-derived language back nearly 2,000 years to the Roman occupation. The nation's oil reserves were once the largest in Europe—they fueled the Nazi war machine—but now appear to be nearing depletion. The economy began to decline in the 1980s because of the repressive and destructive policies of its communist leader, Nicolae Ceausescu. In 1989, the government was overthrown and the hated dictator was executed. In the happier days of the past, the capital, Bucharest, was referred to as the "Paris of the Balkans"— many of its neighborhoods were designed after the French capital. Life in modern Romania has been hardest for over 2 million persecuted Hungarian Magyars, the largest ethnic minority in Europe. Their ancestors were trapped by boundary changes after World War I. Many live in the agricultural and recreational region of Transylvania (home of the fictitious Count Dracula) in the western part of the country. Much of Romania is dominated by the Carpathian Mountains, which trace a wide arc running north to south. Far to the east, where the Danube River reaches the Black Sea, is the largest delta in Europe and home to 300 bird species.

## YUGOSLAVIA

**Area:** 98,842 sq. mi. (255,804 km²). **Population:** 23,400,000. **Capital:** Belgrade, 770,000. **Government:** Republic. **Language:** Serbo-Croatian; Macedonian; Slovenian. **Religion:** Eastern Orthodox 50%; Roman Catholic 40%; Moslem 10%. **Exports:** Engineering products, vehicles, metals, timber, and food. **Climate:** Mediterranean on the coast, continental inland. □ Yugoslavia is being torn from within by the most diverse population in Eastern Europe. After World War I, six historic nations (Bosnia Hercegovina, Croatia, Macedonia, Montenegro, Slovenia, and Serbia) were brought together to form Yugoslavia. Serbia also has two autonomous provinces (Vojvodina, ethnic Hungarians; Kosovo, ethnic Albanians). Under the 35 year rule of Marshal Tito, following World War II, each of the six nations was a socialist republic under a liberal communist government in Belgrade. Tito broke away from the Soviet-controlled Eastern Bloc, aligned Yugoslavia with the Third World, and received Western aid. The nation was opened to tourists, who were especially attracted to the hundreds of islands (the peaks of a sunken range) on the 200 mi. (320 km) Dalmatian coast along the Adriatic Sea. After Tito's death, conflicts between ethnic and national groups increased in frequency and intensity. There were also regional threats of secession and the desire for democratic change. The most serious clashes have occurred between Eastern Orthodox Serbs (8.4 million) and Roman Catholic Croats (4.7 million). The northern republics of Croatia and Slovenia, which have done away with communism, are the most economically advanced and tend to identify more with their northern neighbors than to the rest of Yugoslavia.

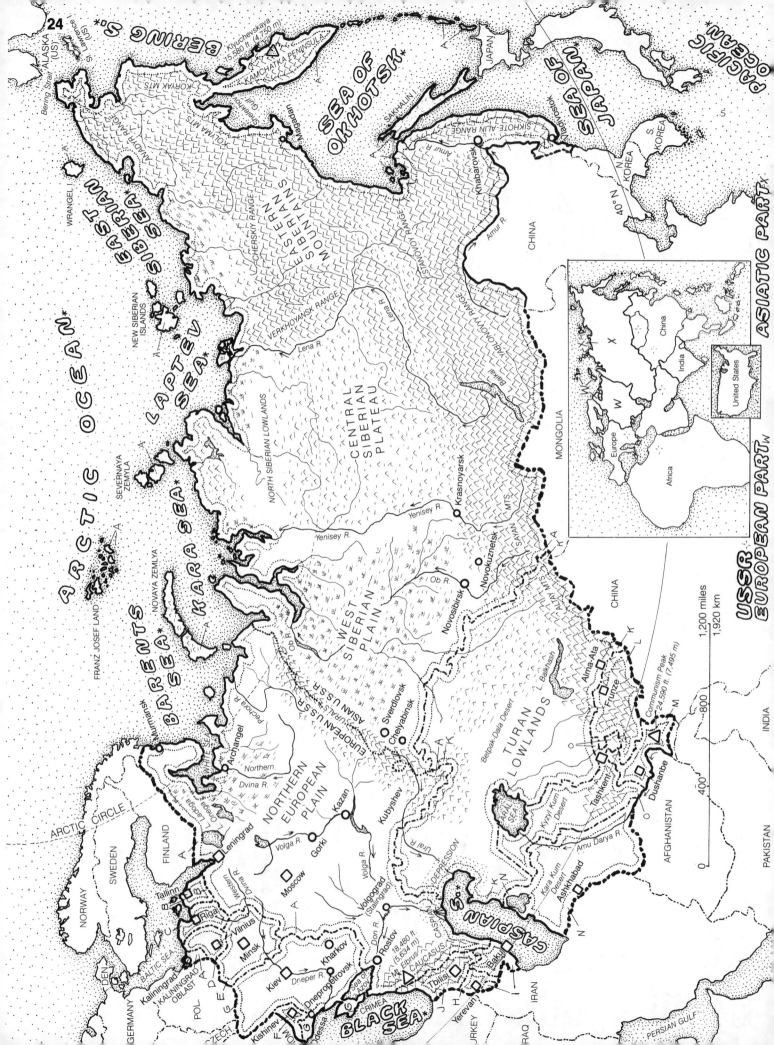

24

ARCTIC OCEAN*

BERING S.*

SEA OF OKHOTSK*

SEA OF JAPAN*

PACIFIC OCEAN*

ALASKA (US)

WRANGEL I.

EASTERN SIBERIA*

WESTERN SIBERIA*

Bering Strait

S. Lawrence I. (US)

KORYAK MTS

ANADYR RANGE

KAMCHATKA PENINSULA

Klyuchevskaya 15,560 ft. (4,750 m)

KOLYMA MTS

CHERSKY RANGE

EASTERN SIBERIAN MOUNTAINS

VERKHOYANSK RANGE

YABLONOVY RANGE

STANOVOI RANGE

SIKHOTE-ALIN RANGE

Magadan

Shelikhov Gulf

SAKHALIN

Amur R.

Khabarovsk

JAPAN

N. KOREA

S. KOREA

CHINA

40° N

Vladivostok

Amur R.

NEW SIBERIAN ISLANDS

LAPTEV SEA*

Lena R.

Lena R.

L. Baikal

CENTRAL SIBERIAN PLATEAU

SEVERNAYA ZEMYLA

KARA SEA*

NORTH SIBERIAN LOWLANDS

Yenisey R.

Yenisey R.

Krasnoyarsk

SAYAN MTS

ALTAI MTS

MONGOLIA

FRANZ JOSEF LAND

BARENTS SEA*

NOVAYA ZEMLYA

Ob R.

WEST SIBERIAN PLAIN

Ob R.

Novokuznetsk

Novosibirsk

CHINA

Murmansk

WHITE SEA

Pechora R.

EUROPEAN USSR / URALS / ASIAN USSR

Sverdlovsk

Chelyabinsk

L. Balkhash

Alma-Ata

Frunze

Communism Peak 24,590 ft. (7,495 m)

TURAN LOWLANDS

Betpak-Dala Desert

ARAL SEA

Kyzyl Kum Desert

Tashkent

Dushanbe

AFGHANISTAN

Archangel

Northern Dvina R.

NORTHERN EUROPEAN PLAIN

Kazan

Kubyshev

Kara Kum Desert

Amu Darya R.

Ashkhabad

Ural R.

ARCTIC CIRCLE

NORWAY

SWEDEN

FINLAND

L. Onega

L. Ladoga

Leningrad

Tallinn

Riga

Vilnius

Minsk

Western Dvina R.

Moscow

Gorki

Volga R.

Volga R.

CASPIAN DEPRESSION

CASPIAN S.*

IRAN

GERMANY

DEN

POL.

CZECH.

Kaliningrad

KALININGRAD OBLAST

BALTIC SEA

Kishinev

Kiev

Kharkov

Dneper R.

Dnepropetrovsk

Odessa

CRIMEA

Sea of Azov

BLACK SEA*

Rostov

Don R.

Volgograd (Stalingrad)

CAUCASUS

M. Elbrus 18,480 ft. (5,634 m)

Tbilisi

Baku

Yerevan

TURKEY

IRAQ

PERSIAN GULF

PAKISTAN

INDIA

USSR

EUROPEAN PART

ASIATIC PART

Europe

Africa

China

India

United States

MONGOLIA

CHINA

1,200 miles
1,920 km

800

400

0

N. EUROPEAN PLAIN P
URAL MOUNTAINS Q
WEST SIBERIAN PLAIN R
CENTRAL SIBERIAN PLATEAU T
N. SIBERIAN LOWLANDS S
TURAN LOWLANDS U
MOUNTAINOUS AREAS V

The USSR is a flat country with tall mountain ranges to the south and east. The Northern European Plain, west of the Urals, is a vast area, with grassy regions (steppes), fertile farmlands, and 75% of the population. The Urals are a long, narrow, low mountain range forming the eastern European border. East of the Urals are various regions of Siberia: the West Siberian Plain, the world's largest flat plain, much of which is either marshy or frozen; the Central Siberian Plateau; the permanently frozen North Siberian Lowlands; and the Turan Lowlands, which contain the Kara Kum and Kyzyl Kum deserts. Southeast of these deserts is the highest point in the country, Communism Peak (24,590 ft, 7,495 m).

Permafrost (permanently frozen ground) covers most of the north. Much of the nation's vast river and canal network is frozen during winter. The sheer size of the USSR and its abundant resources enable it to be the leading producer of oil, natural gas, iron ore, steel, timber, coal, lead, manganese, titanium, mercury, and potash. It is also the number one grower of apples, rye, barley, milk, oats, potatoes, sugar beets, and wheat (which it still must import to meet domestic demand). The USSR has the longest railroad line: the 5,600 mi. (8,960 km) Trans-Siberian Railway connects Moscow with Vladivostok. The country has the most public libraries, the largest fishing fleet, and the largest space program. Eleven cities in the USSR have populations exceeding 1 million. The largest, Moscow, in addition to being the seat of government, is the nation's industrial center. The next largest, the port of Leningrad (4,900,000), was formerly known as St. Petersburg. During Czarist times it was the capital of Russia and the nation's "window to the West." The magnificent buildings and boulevards were designed by western architects in the employ of Peter the Great. During World War II, over 1 million residents died of starvation; the Germans kept the city under siege for nearly 3 years.

The natural features of the USSR are also impressive: it has the world's largest forest, largest flat plain, and largest saltwater lake (the Caspian Sea). Lake Baikal, in Siberia, is the world's deepest freshwater lake—it is over a mile deep and holds as much water as all the Great Lakes of North America combined. Lake Ladoga, near Leningrad, is Europe's largest lake; Mt. Elbrus is Europe's highest peak, and the Volga River is Europe's longest river. East of the Ural Mountains is Siberia, the Asiatic part of the Russian republic. Political prisoners were sent there long before the communists took power. Siberia is similar to 19th-century America's "wild west," and a major effort is being made to encourage migration to this enormous mineral-rich area.

Only 52% of the USSR's diverse population is Russian. The Russians, along with Ukrainians (16%) and Belorussians or White Russians (3.3%), come from a Slavic background, like Poles, Czechs, Yugoslavs, and Bulgarians. People of these three republics have traditionally belonged to the Eastern Orthodox Church (despite the previous official policy restricting religous practice). The western republic of Moldavia has a Roman heritage and language similar to that of Romania, from which it was taken. The Baltic republic of Estonia has a cultural heritage and language similar to that of Finland. Estonians and Latvians (like other northern Europeans) are Protestants. Lithuanians (like their Polish neighbors) are Roman Catholics. People from the Caucasus republic of Azerbaijan and the Central Asian republics of Kazakhstan, Kirghizia, Turkmenistan, and Uzbekistan are Muslims with Turko-Tatar backgrounds. Tadzhikistan is a Persian-speaking republic. The people of the other two Caucasus republics, Georgia and Armenia, are Christians. Their ancient cultures are distinctly different. Armenia was the world's first Christian state (parts of that territory are now in Turkey and Iran). The five Asian republics were annexed in the 19th century by Imperial Russia; all the others were seized by the Soviets, either during a brief period following the 1917 Revolution (1920–1922) or in 1940, when four republics (Moldavia, Estonia, Latvia, and Lithuania) were taken with the collusion of Germany. Moscow shifted in the late 1980s toward a more free and democratic Soviet Union; this uncorked a

passionate desire for independence in many republics. Ethnic and religious hatreds also emerged and violent encounters ensued, particularly between Armenians and Azerbaijanis. Many republics, including Russia, have recently declared their independence of the Soviet Union. It seems unlikely that the USSR will continue to exist under its present political structure. It is possible that a confederation of independent states, similar to the British Commonwealth, is in store for the Soviet Union.

The 1917 revolution, led by Vladimir I. Lenin, which brought the communist party to power, marked the beginning of the USSR's remarkable change from a poor, illiterate agricultural country to its current international status. The death of Lenin in 1924 brought to power Joseph Stalin, who instituted a police state that outlived his death in 1953. During and after his reign, the Soviets made sizable advances in the industrial and military sector. The nation is close to the US in military strength, and until it was overtaken by Japan in 1989, it was the world's second largest industrial power. But the Soviet people have paid an enormous price for this accomplishment. Stalin was responsible for the death of millions in the restructuring of Soviet society. Added to those atrocities were the incredible destruction and loss of life during World War II. Though hunger, disease, unemployment, and illiteracy have largely been eliminated, its very low standard of living places the Soviet Union on the economic level of a Third World nation, albeit a very powerful one. Housing and consumer goods have been of poor construction and always in short supply. The economy has been managed as if the nation were in a permanent state of war, with industrial or military demands receiving first claim on available resources. Food has never been plentiful, partly because of an unfavorable climate, but mostly because of communism's failure to motivate farmers. There was widespread opposition to the collectivization of small farms into large, centrally managed organizations. When farmers were finally allowed to set aside private plots for personal use and profit, production increased.

In the 1980s, under the leadership of Mikhail Gorbachev, the Soviet Union recognized that its secretive, controlled economy was falling farther behind the dynamic capitalist nations. Gorbachev undertook a massive restructuring of Soviet society ("perestroika") with an emphasis on openness ("glasnost") in the conduct of government, elections, and personal expression. Attempts to infuse free-market techniques into the economy at first met with considerable resistance. Though the restoration of political, human, and religious liberties was eagerly received, 70 years of communism bred a society that is uncertain and fearful of a free economic system that carries with it the likelihood of rising prices, unemployment, and a competitive spirit.

CN: Save your lightest colors for P–V. (1) When coloring the Republic of Russia (A) note the detached part (Kaliningrad Oblast) next to Poland. (2) Except for two Baltic islands that belong to Estonia (B), all islands with dark outlines belong to Russia (A). (3) The squares and larger circles shown on this map represent cities of over 1 million. The smaller circles represent cities of over 250,000.

RUSSIAN REPUBLIC A MOSCOW

BALTIC REPUBLICS :
ESTONIA B TALLINN
LATVIA C RIGA
LITHUANIA D VILNIUS

WESTERN REPUBLICS :
BELORUSSIA E MINSK
MOLDAVIA F KISHINEV
UKRAINE G KIEV

CAUCASUS REPUBLICS :
ARMENIA H YEREVAN
AZERBAIJAN I BAKU
GEORGIA J TBILISI

ASIAN REPUBLICS :
KAZAKHSTAN K ALMA-ATA
KIRGHIZIA L FRUNZE
TADZHIKISTAN M DUSHANBE
TURKMENISTAN N ASHKHABAD
UZBEKISTAN O TASHKENT

Area: 8,649,500 sq. mi. (22,400,200 km²). Population: 275,000,000. Capital: Moscow, 7,700,000. Government: Republic. Language: Russian (official); 100 dialects. Religion: Russian Orthodox; Islam. Exports: Oil, coal, iron ore, machinery, wood, furs, and caviar. Climate: Baltic coast is moderate; Black and Caspian Sea regions are subtropical; the rest of the country has extreme continental weather with long winters. □ The Union of Soviet Socialist Republics (USSR), or the Soviet Union, is often referred to as "Russia," but the Russian Soviet Federal Socialist Republic is only one of 15 republics. Russia is by far the largest; it encompasses 75% of the nation's land area. It has 52% of the population and is the most politically influential; the seat of government, the Kremlin, is located in Moscow, the capital of both Russia and the USSR. The 14 other republics are located on Russia's western and southwestern borders. Russian is the official language, but it is usually the second language for the populations of the other republics. Each republic represents a major ethnic group with its own language and culture, and some republics include many minorities. There are about 100 different nationalities with their own languages in the Soviet Union.

The USSR is the world's largest country and has the third largest population. It is 2.5 times the size of the United States and occupies nearly a sixth of the earth's landmass. Because of its northerly location and its distance from the warming influence of the Atlantic and Pacific Oceans, most of the USSR experiences long and frigid winters. Some of the earth's coldest temperatures occur in northeastern Siberia (the record is -90° F, -68° C).

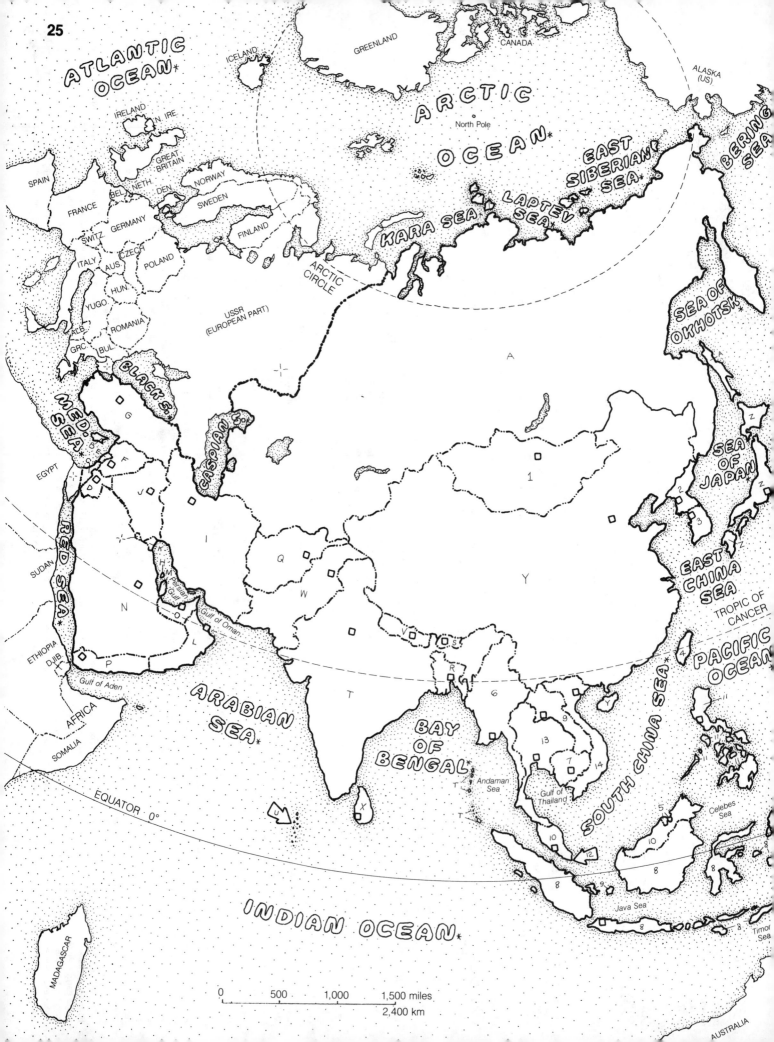

# 25

# ASIA: THE COUNTRIES

CN: (1) On the small map, use gray for Asia in the dark outline (including the islands). (2) On the large map, color the two arrows (near the equator) representing the locations of the Maldives (U) and Singapore (12).

## USSR A (ASIAN PART)

## MIDDLE EAST I
CYPRUS B / NICOSIA
ISRAEL C / JERUSALEM
JORDAN D / AMMAN
LEBANON E / BEIRUT
SYRIA F / DAMASCUS
TURKEY G / ANKARA

## MIDDLE EAST II
BAHRAIN H / MANAMA
IRAN I / TEHRAN
IRAQ J / BAGHDAD
KUWAIT K / KUWAIT CITY
OMAN L / MUSCAT
QATAR M / DOHA
SAUDI ARABIA N / RIYADH
UNITED ARAB EMIRATES O / ABU DHABI
YEMEN P / SANA

## SOUTH
AFGHANISTAN Q / KABUL
BANGLADESH R / DHAKA
BHUTAN S / THIMPHU
INDIA T / NEW DEHLI
MALDIVES U / MALE
NEPAL V / KATHMANDU
PAKISTAN W / ISLAMABAD
SRI LANKA X / COLOMBO

## FAR EAST
CHINA Y / BEIJING
JAPAN Z / TOKYO
MONGOLIA 1 / ULAN BATOR
NORTH KOREA 2 / PYONGYANG
SOUTH KOREA 3 / SEOUL
TAIWAN 4 / TAIPEI

## SOUTHEAST
BRUNEI 5 / BANDAR SERI BEGAWAN
BURMA 6 / RANGOON
CAMBODIA 7 / PHNOM PENH
INDONESIA 8 / JAKARTA
LAOS 9 / VIENTIANE
MALAYSIA 10 / KUALA LUMPUR
PHILIPPINES 11 / MANILA
SINGAPORE 12 / SINGAPORE
THAILAND 13 / BANGKOK
VIETNAM 14 / HANOI

Asia, the largest continent (17,230,000 sq. mi., 44,625,700 km²), contains 30% of the earth's landmass. It is both the widest and the deepest continent: it stretchs 6,000 mi. (9,600 km) from Turkey's Aegean coastline to the Pacific shores of Japan and covers a similar distance from Siberia's Arctic tundra to the tropical Indonesian Islands south of the equator. Asia has the most people (3,200,000,000)—nearly 60% of the world's population. One out of every three human beings lives in either China or India. Because so much of Asia is extremely dry or mountainous, the world's most crowded population centers are generally along the continent's coastlines or river valleys.

Asia is separated from Europe by an imaginary line that passes down the Ural Mountains of the Soviet Union, south to the Caspian Sea, then westward along the slopes of the southern Caucasus to the Black Sea. Asia was connected to Africa until the Suez Canal was built. The presence of the canal places the Sinai Peninsula, a part of the African nation of Egypt, within Asia. Asia was also connected to North America, but when the ocean levels rose after the last ice age, the 50 mi. (80 km) Bering Strait was created, separating Siberia (USSR) and Alaska (US).

Siberia is the Asian part of the Soviet Union's Russian republic. Siberia and the five Asian republics of the USSR occupy the northern half of Asia. Central Asia (Mongolia and northern China) is a sparsely populated region in which animal herding is the principal industry. Most Asian nations are poor and predominantly agricultural. In the Far East, Japan and the "Little Dragons" (Hong Kong, Singapore, South Korea, and Taiwan) have become extraordinarily productive industrial powers. The wealthiest nations (per capita) in the world are located on the Arabian Peninsula. These desert monarchies are sitting on well over half of the world's known oil reserves.

Asia was the birthplace of many of the world's oldest civilizations. The Tigris-Euphrates Valley of the Middle East, the Indus River Valley of Pakistan, and the Huang He River Valley of China were the locations of flourishing, advanced societies. All the world's major religions originated in Asia. Judaism, Christianity, and Islam came from the Middle East; Hinduism, Buddhism, Confucianism, Taoism, and Shintoism originated in South Asia and the Far East. Christianity has the most adherents worldwide, but it plays a minor role in Asia. Hinduism, in India and Nepal, has the most followers in this part of the world. Islam is the second largest religion in Asia; it is the dominant faith in Pakistan, Bangladesh, Malaysia, and Indonesia as well as Turkey, Iraq, Iran, and the Arab nations of the Middle East.

Asians generally fall into the Caucasoid and Mongoloid racial groups (see Plate 46). Caucasoids include the people of the Middle East (Arab countries, Israel, Turkey, Iraq, Iran, and Afghanistan) and the Indians of South Asia (India, Pakistan, and Bangladesh). The Mongoloid race includes the people of the Far East and Southeast Asia.

# ASIA: THE PHYSICAL LAND

## PRINCIPAL RIVERS:

AMU DARYA A
AMUR B
BRAHMAPUTRA C
EUPHRATES D
GANGES E
HUANG HE (YELLOW) F
INDUS G
IRRAWADDY H
IRTYSH I
LENA J
MEKONG K
OB L
SALWEEN M
TIGRIS N
XI JIANG (PEARL) O
YANGTZE P
YENISEY Q

## PRINCIPAL MOUNTAIN RANGES:

ALTAI R
ELBURZ S
HIMALAYAS T
HINDU KUSH U
KUNLUN V
PAMIRS W
TIAN SHAN X
URALS Y

## LAND REGIONS:

DESERT Z
RAIN FOREST 1
LOWLANDS 2
MOUNTAINS 3
PLATEAU 4
HIGHLANDS 5
TUNDRA 6

Asia has the highest and lowest points on the planet (Mt. Everest and the Dead Sea). Asia is the continent with the tallest mountains, the largest and highest plateaus, the largest deserts, most of the longest rivers, the deepest lake, the largest forest region, the largest flat plains, the most active volcanoes, the most earthquakes, and the hottest, coldest, driest, and wettest climates. This is an impressive list, but much of Asia amounts to a vast, inhospitable environment where few people live.

The most northerly part of Asia is the Arctic portion of Siberia (USSR). South of this tundra is a huge highlands region containing the world's largest coniferous forest. To the west is the world's largest, flattest lowlands, the West Siberian Plain. South of these regions, and spanning the midsection of Asia from Saudi Arabia to southern Mongolia, is a succession of large deserts, barren plateaus, and rugged mountains. Unlike the mountains of North and South America and Europe, Asia's massive ranges are located in the center of the continent. Surrounded by some of the tallest peaks is the Plateau of Tibet—the "Roof of the World" (15,000 ft., 4,573 m) and the highest inhabited plateau. Along Tibet's southern border are the Himalayas, the world's tallest mountains, including the tallest peak, Mt. Everest (29,028 ft., 8,848 m). These geologically young mountains continue to grow as the tectonic plate carrying the subcontinent of India grinds under the Eurasian plate (see Plate 2). The Asian subcontinent (the Indian Peninsula) is a triangular plateau that is subject to heavy monsoonal rainfall. Southeast Asia includes the rain forests of the Indochina Peninsula, the Malay Peninsula and Archipelago, and the Philippine Islands. Earthquake-prone regions—the Malay Archipelago, the Philippines, and the islands of Japan—are the peaks of seismically active sunken mountain ranges. The Indonesian island of Java has the greatest concentration of active volcanoes in the world.

The central mountain ranges of Asia contain the headwaters for many of the world's longest rivers. The Yangtze (3,915 mi., 6,265 km) is the third longest (after the Nile and Amazon) and is China's most important river.

Because Asia is so vast, much of its interior is far from the moderating influence of the Atlantic and the Pacific; it is subject to the earth's greatest temperature extremes. The brutal temperatures and extremely dry climate of Central Asia create the kind of weather that only nomadic herders will endure. Southwest Asia (the Middle East) is both very dry and very hot—summer temperatures are often in excess of 120° F (49° C). South and Southeast Asia are not quite as hot, but high humidity and torrential rainfall can make the climate equally uncomfortable. The summer monsoon winds bring heavy, ocean-bred rainstorms that make these regions such successful rice producers.

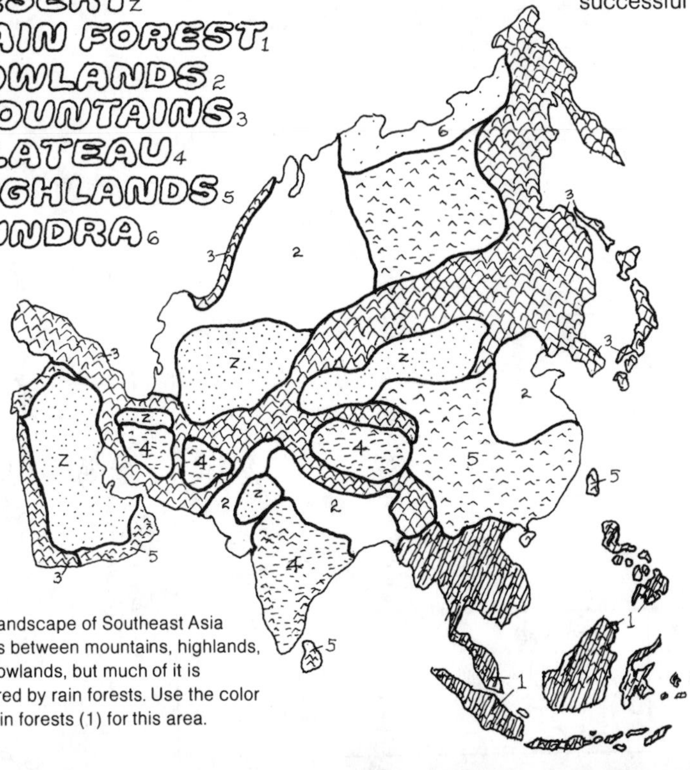

The landscape of Southeast Asia varies between mountains, highlands, and lowlands, but much of it is covered by rain forests. Use the color for rain forests (1) for this area.

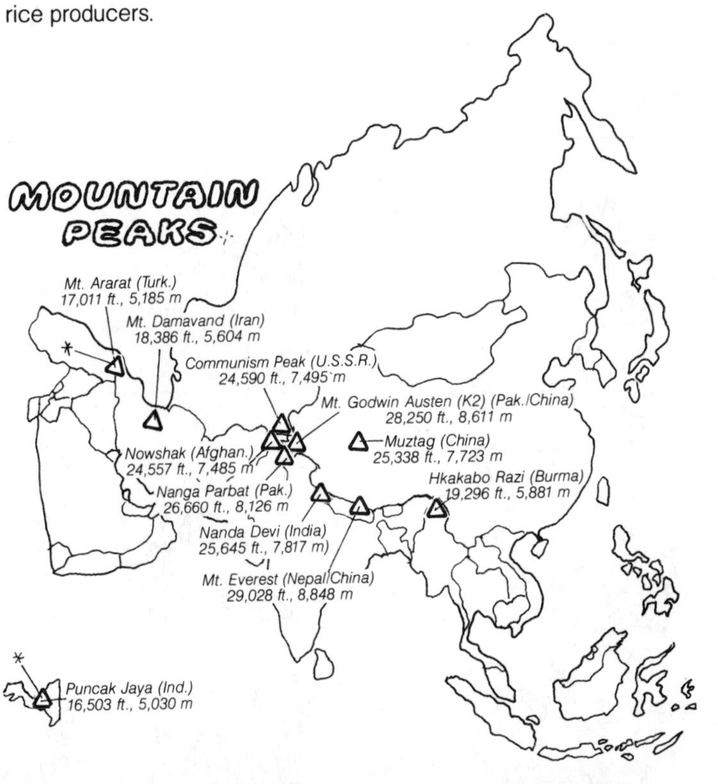

## MOUNTAIN PEAKS:

Mt. Ararat (Turk.)
17,011 ft., 5,185 m

Mt. Damavand (Iran)
18,386 ft., 5,604 m

Communism Peak (U.S.S.R.)
24,590 ft., 7,495 m

Mt. Godwin Austen (K2) (Pak./China)
28,250 ft., 8,611 m

Nowshak (Afghan.)
24,557 ft., 7,485 m

Muztag (China)
25,338 ft., 7,723 m

Nanga Parbat (Pak.)
26,660 ft., 8,126 m

Hkakabo Razi (Burma)
19,296 ft., 5,881 m

Nanda Devi (India)
25,645 ft., 7,817 m

Mt. Everest (Nepal/China)
29,028 ft., 8,848 m

Puncak Jaya (Ind.)
16,503 ft., 5,030 m

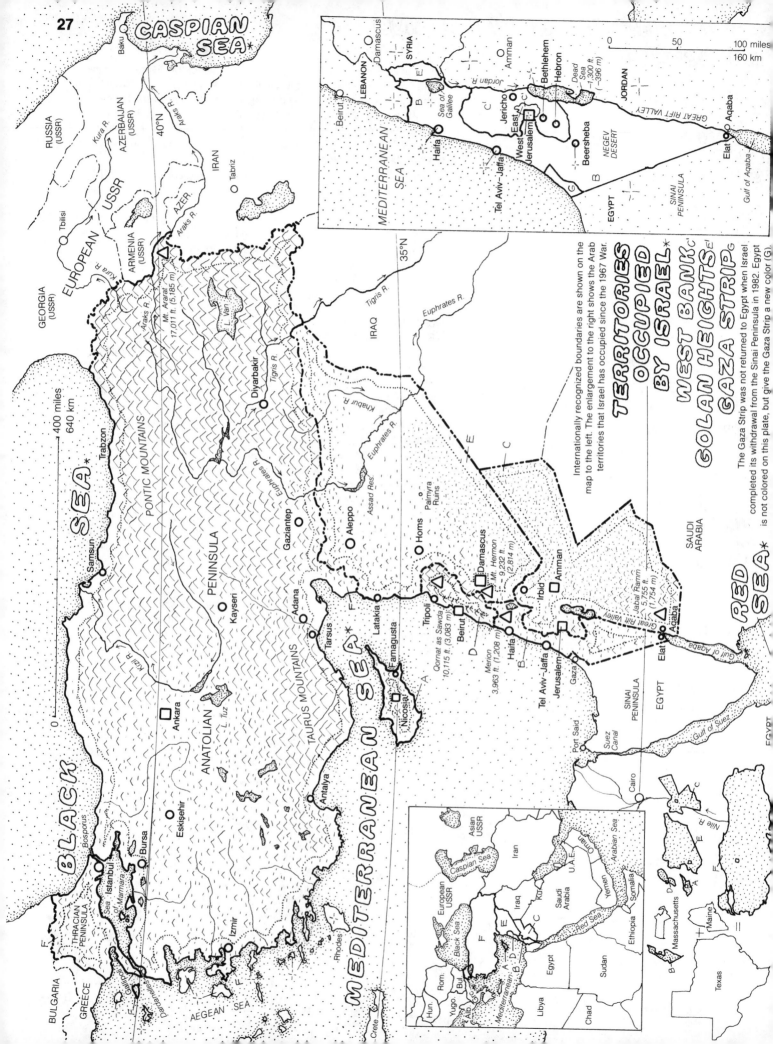

27

CASPIAN SEA*

Damascus
SYRIA
LEBANON
E'
B'
D'
Beirut
Sea of Galilee
Jordan R.
C'
Amman
Bethlehem
Hebron
West Jerusalem
East Jerusalem
Jericho
Dead Sea −1,300 ft. (−396 m)
JORDAN
Haifa
Tel Aviv–Jaffa
C'
B'
Beersheba
NEGEV DESERT
B
G
EGYPT
SINAI PENINSULA
Elat
Gulf of Aqaba
MEDITERRANEAN SEA

0    50    100 miles
0    160 km

RUSSIA (USSR)
Baku
AZERBAIJAN (USSR)
Kura R.
Tbilisi
40°N
GEORGIA (USSR)
EUROPEAN
USSR
ARMENIA (USSR)
AZER.
Araks R.
IRAN
Tabriz
Mt. Ararat 17,011 ft. (5,185 m)
L. Van
Diyarbakir
Tigris R.
35°N
Tigris R.
IRAQ
Euphrates R.
Khabur R.
Euphrates R.
Euphrates R.
Assad Res.
PONTIC MOUNTAINS
Trabzon
BLACK SEA*
Samsun
PENINSULA
ANATOLIAN
Kayseri
Ankara
Eskişehir
L. Tuz
Kızıl R.
Gaziantep
Aleppo
Homs
Palmyra Ruins
Damascus
Mt. Hermon 9,232 ft. (2,814 m)
Adana
Tarsus
TAURUS MOUNTAINS
Latakia
Tripoli
Qornat as Sawda 10,115 ft. (3,083 m)
Beirut
Meiron 3,963 ft. (1,208 m)
Haifa
D
B
Tel Aviv–Jaffa
Jerusalem
Gaza
Irbid
Amman
Great Rift Valley
Jabal Ramm 5,755 ft. (1,754 m)
SAUDI ARABIA
Elat
Aqaba
Gulf of Aqaba
RED SEA*
Antalya
Famagusta
Nicosia
Rhodes
Izmir
Crete
MEDITERRANEAN SEA*
AEGEAN SEA
GREECE
BULGARIA
THRACIAN PENINSULA
Gallipoli
Dardanelles
Bosporus
Istanbul
Sea of Marmara
Bursa
Gulf of Suez
Port Said
Suez Canal
Cairo
EGYPT
SINAI PENINSULA
Nile R.
EGYPT

400 miles
640 km

Internationally recognized boundaries are shown on the map to the left. The enlargement to the right shows the Arab territories that Israel has occupied since the 1967 War.

TERRITORIES OCCUPIED BY ISRAEL*
WEST BANK
GOLAN HEIGHTS
GAZA STRIP

The Gaza Strip was not returned to Egypt when Israel completed its withdrawal from the Sinai Peninsula in 1982. Egypt is not colored on this plate, but give the Gaza Strip a new color (G).

Asian USSR
Caspian Sea
Iran
Oman
U.A.E.
Arabian Sea
European USSR
Black Sea
Iraq
Kuwait
Saudi Arabia
Yemen
Somalia
Mediterranean Sea
Egypt
Red Sea
Sudan
Ethiopia
Chad
Libya
Rom.
Hun.
Yugo.
Bul.
Alb.
Gr.
Massachusetts
Maine
Texas

# ASIA: MIDDLE EAST I

Except for Cyprus, Iran, Israel, and Turkey, the countries of the Middle East are Arab—their people speak Arabic and follow the Islamic religion. Citizens of Iran and Turkey are not Arabs, but they too are Muslims (followers of Islam). Most Cypriots are Greek Orthodox Christians, and most Israelis are Jewish.

The part of the Middle East shown on this map does not have the oil reserves that the modern world so desperately needs, but it does have great religious significance for Jews, Christians, and Muslims. Judaism and Christianity were born here, and the city of Jerusalem is holy for Muslims as well. Today, the region is marked by a constant state of tension punctuated by periodic conflicts. It is a case of Muslim versus Jew (the Arab nations and Israel remain in a technical state of war); Christian versus Muslim (in Lebanon and Cyprus); and Muslim versus Muslim (Islamic factions clash in Lebanon).

Even the earth is divided in the Middle East: the Great Rift Valley, a break in the earth's crust, runs from Syria south under the Red Sea (Plate 35). It defines the border of Israel and Jordan and holds the Jordan River, the Sea of Galilee, and the Dead Sea. The latter is the saltiest body of water on earth (28% salt vs. 3.5% in seawater), and it is the lowest point on the earth's surface (−1,300 ft., −396 m). The sea's absence of life and absolute silence suggest no other name.

## CYPRUS A

**Area:** 3,572 sq. mi. (9,251 km²). **Population:** 700,000. **Capital:** Nicosia, 162,000. **Government:** Republic. **Language:** Greek; Turkish. **Religion:** Greek Orthodox; Islam. **Exports:** Citrus, copper, wine, and food products. **Climate:** Mediterranean. □ The words "Cyprus" and "copper" come from the Greek "kypros." Today agriculture, not copper, is the principal industry on this ruggedly beautiful island with an ancient past. The British Empire, the last of its many rulers, granted independence in 1960. A new constitution provided for joint rule between the Greek Orthodox majority (77%) and the Turkish Muslim minority, but their differences could not be resolved and fighting broke out. It persisted until an invasion in 1974 by soldiers from Turkey. A cease-fire was mediated by the United Nations and a line was drawn across northern Cyprus and through Nicosia, the 5,000-year-old capital. The Turkish minority were given control of the northern third of the island, which they and Turkey consider to be an independent nation. The rest of the world regards Cyprus as a single nation under a Greek Cypriot government.

## ISRAEL B

**Area:** 8,018 sq. mi. (20,767 km²). **Population:** 4,500,000. **Capital:** Jerusalem, 475,000. **Government:** Republic. **Language:** Hebrew; Arabic. **Religion:** Judaism; Islam. **Exports:** Oranges, produce, polished diamonds, and manufactured goods. **Climate:** Mediterranean on the coast, desert inland. □ A fifth of the world's Jews live in this land from which their ancestors were driven by repeated invasions, 2,000 years ago. For centuries, Jews scattered around the world have dreamed of returning to this land. Hebrew, the language of the Bible, has been resurrected as Israel's official language. In the late 19th century, European Jews formed the Zionist movement to create a homeland in the Middle East. After World War I, Britain, which received a mandate from the League of Nations to govern Palestine, permitted the gradual immigration of European Jews. But Palestinian Arabs, whose people had lived there for 1,500 years, were alarmed at the growing number of immigrants and persuaded Britain to restrict immigration. After World War II, pressure to immigrate mounted as survivors of the Holocaust demanded entry. The United Nations carved Palestine into two states, one Jewish, and one Arab, with the city of Jerusalem under international control. The Arab nations rejected this plan and attacked the newly created nation in 1948. The Jews won the war and seized half of what was intended to be the Arab state, absorbing many Palestinians in the process. Thousands of other Palestinians fled to neighboring Arab countries to escape the fighting. Afterward, they were denied reentry by the Israelis, and today, those Palestinians and their descendants make up the 2,500,000 permanently displaced refugees living in Lebanon, Jordan, and Syria. The Arabs (and their descendents) who did not leave Israel during the war number around 750,000 (15% of the population) and maintain their own culture. They have a higher standard of living than most Arabs in the Middle East and are accorded most of the same rights as other Israeli citizens, but they are often discriminated against and live as a segregated minority. Forty years and three wars after its founding, Israel had acquired additional Arab territories: the Gaza Strip from Egypt, the Golan Heights from Syria, the West Bank and East Jerusalem from Jordan. In 1987, a resistance movement (intifada) was formed among the million Palestinians in the Gaza Strip and the West Bank to protest the Israeli occupation. Israel has been widely criticized for responding with excessive force and also for allowing Jewish settlers to move into the occupied West Bank of the Jordan River. Israel is technically at war with all its Arab neighbors except Egypt; the two countries signed a peace treaty in 1979 in which Israel returned the Sinai Peninsula but not the Gaza Strip to Egypt.

The capital of Israel is the modern city of West Jerusalem. When Israel captured and formally annexed East Jerusalem, it included the Old City, the site of ancient Jerusalem, and Jews were once again able to pray at the "Wailing Wall" believed to be part of the ancient Temple of Solomon. The old city is holy to Christians and Muslims as well. It was the setting for many events in the life of Christ, and Muslims believe that it was there that Muhammed ascended to Heaven, making Jerusalem nearly as sacred to them as Mecca and Medina. The modern cities of Tel Aviv-Jaffa and Haifa are centers of industry and commerce. Out of the desert, Israel has created the most productive farmland in the Middle East. This tiny nation of less than 5 million is also one of the world's most militarily powerful and technologically advanced societies. Much of its strength is the result of massive aid from the United States, reparations from Germany (for crimes against the Jewish people), and support from Jews around the world.

## JORDAN C

**Area:** 37,600 sq. mi. (97,384 km²). **Population:** 4,100,000. **Capital:** Amman, 1,250,000. **Government:** Monarchy. **Language:** Arabic. **Exports:** Phosphates, produce, tobacco, and oil products. **Climate:** Hot desert. □ Except for 15 mi. (24 km) of coastline on the Gulf of Aqaba, Jordan is land-locked. About 80% of its land is desert. The nation has no historic basis: as part of the breakup of the Ottoman Empire after World War I, the British made Jordan and other former colonies in the Middle East independent countries. Jordan's only fertile land is in the Jordan River Valley; more than half of this area (the West Bank) and a quarter of Jordan's population were lost to Israel in the Six-Day War of 1967. Jordan wants the West Bank to become a new Palestinian homeland. Israel argues that because Palestinians are the majority in Jordan, it is already a Palestinian state. Many Jordanians are members of Bedouin tribes that no longer live the life of nomads. Amman, the capital, has replaced war-torn Beirut, Lebanon, as the financial center for this part of the Middle East.

## LEBANON D

**Area:** 4,024 sq. mi. (10,422 km²). **Population:** 3,600,000. **Capital:** Beirut, 750,000. **Government:** Republic. **Language:** Arabic. **Religion:** Islam; Roman Catholic. **Exports:** Tobacco, fruits and vegetables, and small industrial products. **Climate:** Mild, but varies according to terrain. □ Prior to the 1970s, beautiful Lebanon was the most stable and prosperous nation in this region. Its capital, Beirut, was called the "Paris of the Middle East." It was the regional center for finance and commerce and it provided shipping facilities for Jordan, Iraq, and Syria. Since then the country has been in an almost perpetual state of civil war, with Christians fighting Muslims and Muslims fighting among themselves. The Maronites in Lebanon represent the highest percentage (40%) of Christians in any Arab country. An Arab majority of Druse, Shiite, and Sunni Muslims have been contesting the traditional power wielded by the Roman Catholic Maronites. The urban warfare of Beirut has spilled over into the countryside. Before the war, a visitor traveling the 40 mi. (64 km) width of the country could see Mediterranean beaches, a green coastal plain, snowcapped mountains (in the winter), the Beqaa Valley, and the edge of the Syrian desert. The famed cedars of Lebanon are nearly extinct, but reforestation programs are intended to bring back the magnificent trees.

## SYRIA E

**Area:** 71,510 sq. mi. (185,211 km²). **Population:** 10,800,000. **Capital:** Damascus, 1,200,000. **Government:** Socialist republic. **Language:** Arabic. **Religion:** Islam. **Exports:** Cotton, textiles, tobacco, fruits, and oil. **Climate:** The coast is mild; interior is dry. □ Syria is a land filled with the history of ancient civilizations. It was the trade center of the caravan routes that linked the continents of Europe, Asia, and Africa. This area makes up a large part of a river-irrigated region that has sustained life in the Middle East for thousands of years. This region, called the "Fertile Crescent" because of its shape, is bounded by the Tigris and Euphrates Rivers. It begins at the Persian Gulf and arcs northwestward through most of Syria and down into Israel. Damascus, the 5,000 year-old capital of Syria, may be the world's oldest continuously inhabited city. Archaeologists have been excavating the ruins of Palmyra for over 50 years but are far from completely uncovering the ancient trade center. During this century, Syria (like Lebanon) was part of the Ottoman Empire and became a French mandate following World War I. It received independence in 1946. A succession of military regimes have been in power since then. Modest oil deposits help pay for basic government expenses. National animosity focuses on Israel, which continues to occupy the strategic Golan Heights that loom over Israel's northeastern border.

## TURKEY F

**Area:** 300,960 sq. mi. (779,486 km²). **Population:** 55,000,000. **Capital:** Ankara, 1,900,000. **Government:** Republic. **Language:** Turkish. **Exports:** Cotton, carpets, fruit and nuts, tobacco, and chromium. **Climate:** Thrace and the coastal regions are Mediterranean; interior is continental. □ Turkey is made up of two peninsulas, on different continents. Three percent of Turkey is the Thracian Peninsula (Thrace) on the European continent, separated from Turkey proper by the Sea of Marmara and two narrow straits, the Dardanelles and the Bosporus. The remainder of Turkey occupies the Anatolian Peninsula, also called Anatolia or Asia Minor. It is a rugged plateau bordered by mountains on the north, south, and east. The Bosporus divides Turkey's largest city, Istanbul (2,700,000), into European and Asian sections. The city was inhabited long before the ancient Greeks named it Byzantium. The Romans made it the capital of their Empire in 324 AD and changed its name to Constantinople. When it became the capital of the Ottoman Empire in 1453, the Turks gave it its present name. Mount Ararat (17,011 ft., 5,186 m), the tallest peak in the Middle East, is on the eastern border with the USSR and Iran; it is believed by some to be the resting place of Noah's Ark. Both the Euphrates and the Tigris Rivers originate in Turkey. There is concern in Syria and Iraq that Turkey's ambitious hydroelectric and irrigation projects will reduce the flow of those vital rivers. For over 500 years Turkey was the heart of the Ottoman Empire, which ruled the Middle East, North Africa, and southeastern Europe. The empire's collapse came when it fought on the losing German side in World War I. Turkey's present borders were defined in 1923, the year it became a republic. Its first president, Kemal Ataturk, completely westernized the nation. He eliminated traditional modes of dress (including the veil and fez), the Arabic alphabet, Islamic schools and laws, polygamy, and the subjugation of women. With the rise of Islamic fundamentalism in the 1980s, there has been some reversion to the customs of the past.

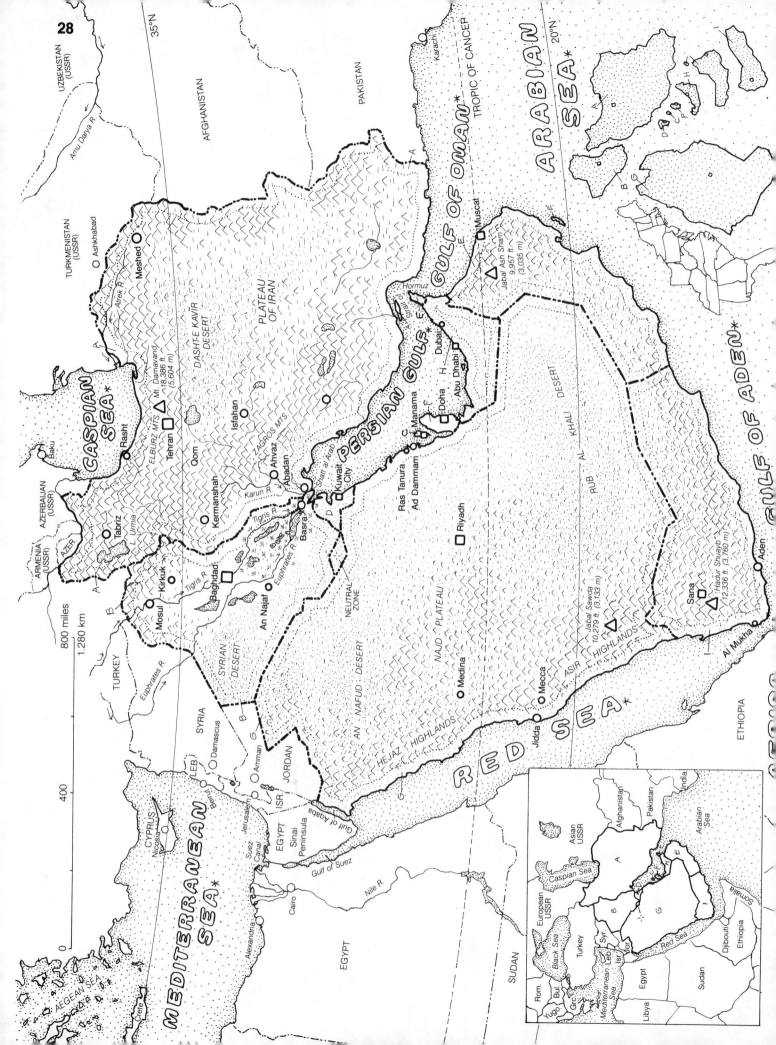

28

After World War II, oil became the major source of revenue for most of these nations. In the 1970s, newly formed OPEC (Organization of Petroleum Exporting Countries) curtailed production and prices skyrocketed. Almost overnight the oil-producing countries of this region went from poor desert nations to rich, urban, welfare states with no taxes and free social and educational services. Because Iran and Iraq spent the 1980s locked in a devastating war, they have not fared as well. Oil prices tumbled after the 1970s, but remain higher than they were before the OPEC nations took control of the ownership and pricing of their reserves. Most Arabian nations earn additional income from worldwide and domestic investments.

The Arabian Peninsula is one of the world's hottest and driest regions. There are virtually no rivers or streams, but underground water surfaces at oases and private wells. Many of these nations can easily afford to distill seawater. The peninsula's barren land slopes eastward from highlands along the Red Sea to the oil-rich lowlands along the Persian Gulf. Most of the Arabian nations became independent of Great Britain in the mid-20th century, but English is still learned as a second language and is spoken in business.

Kurdistan, a region of eastern Turkey, northern Iraq, northwestern Iran, the southern USSR, and the tip of Syria, is a "borderless nation" that seeks a political identity. It is the home of the Kurds, a fiercely independent mountain group of Muslims whose population is estimated at 20 million (the largest ethnic minority in the Middle East). They have resisted all attempts to assimilate them, and a radical group of Kurds in Turkey is currently fighting for independence. After Iraq signed a cease-fire with Iran, it used poison gas against elements of its Kurdish population for siding with the enemy.

## IRAN ☆ A

**Area:** 636,259 sq. mi. (1,648,004 km²). **Population:** 51,000,000. **Capital:** Tehran, 5,800,000. **Government:** Islamic Republic. **Language:** Persian (Farsi). **Religion:** Islam. **Exports:** Oil and oil products, natural gas, cotton, caviar, and Persian rugs. **Climate:** Except for Caspian coast, dry and continental. ☐ The world's top oil exporter. Iran (ih ran') with the profits from enormous oil reserves; this led to his downfall in 1979. Religious conservatives led by the Ayatollah Khomeini came to power and proceeded to run Iran under Islamic law. Iran isolated itself from the world community and received little help when Iraq invaded. An eight-year bloody war ended in a stalemate in 1988. Iran (formerly Persia) rests on a huge plateau with an average elevation of 4,000 ft. (1,220 m), bordered by northern and western mountain ranges that form a giant V that points at Turkey. The northern slope of the Elburz Mountains, facing the Caspian Sea, is the only region with enough rainfall to support agriculture. The Caspian Sea is a rich provider of the sturgeon eggs that support an important caviar industry. On the southern slopes of the Elburz Mountains is the capital, Tehran, the Middle East's largest city and the cultural, religious, and business center of Iran.

## IRAQ ☆ B

**Area:** 169,930 sq. mi. (438,419 km²). **Population:** 17,000,000. **Capital:** Baghdad, 3,300,000. **Government:** Republic (socialist). **Language:** Arabic, Kurdish. **Religion:** Islam. **Exports:** Oil, dates, copper, wool, and hides. **Climate:** Very hot summers and mild winters. ☐ Iraq (ih rak') produces 80% of the world's supply of dates, but oil is by far its major source of revenue. The fertile land between the Euphrates and Tigris rivers, which in ancient times was called Mesopotamia (Greek for "between the rivers"), supported the

earliest recorded civilizations; artifacts date back some 9,000 years. The early Sumerian, Assyrian, and Babylonian societies were founded here. The plow, the wheel, and writing are believed to have originated in this eastern half of the Fertile Crescent. The Tigris River runs through the capital, Baghdad, once the heart of the Arab Empire and a 9th-century center of learning. It was later ravaged by Mongol invaders, and never regained its stature. In 1980, an international crisis was created when Saddam Hussein sent the Iraqi army into Kuwait and declared the oil-rich nation to be part of Iraq. The US and other countries dispatched military forces to Saudi Arabia, and the United Nations voted to embargo all trade into and out of Iraq.

# ARABIAN PENINSULA ☆

## BAHRAIN ☆ C

**Area:** 250 sq. mi. (663 km²). **Population:** 500,000. **Capital:** Manama, 100,000. **Government:** Sheikdom. **Language:** Arabic. **Religion:** Islam. **Exports:** Oil and oil products. **Climate:** Hot summers; warm winters. ☐ This tiny country consists of a 33-island archipelago in the Persian Gulf. Bahrain is the name of the largest island, which is connected to Saudi Arabia by a causeway. Bahrain derives its fortune from oil reserves and by refining Saudi Arabian oil in huge facilities. It has become a Middle Eastern banking and finance center and a headquarters for international corporations. The ruler of this sheikdom, called an emir, controls the cabinet that governs the country.

## KUWAIT ☆ D

**Area:** 6,870 sq. mi. (17,793 km²). **Population:** 1,950,000. **Capital:** Kuwait City, 450,000. **Government:** Emirate. **Language:** Arabic. **Religion:** Islam. **Exports:** Oil and natural gas. **Climate:** Hot summers. ☐ Kuwait, wedged into the northwest corner of the Persian Gulf, may have as much as 10–15% of the world's oil reserves. The wealthy Kuwaiti citizens are outnumbered by a Muslim labor force from India, Iran, and Pakistan and a quarter of a million Palestinian refugees. In 1990, a dispute over border lands, oil production, and access to the Gulf led to the invasion by Iraq. Kuwait's ruling emir fled the country.

## OMAN ☆ E

**Area:** 82,025 sq. mi. (212,445 km²). **Population:** 1,600,000. **Capital:** Muscat, 60,000. **Government:** Sultanate. **Language:** Arabic. **Religion:** Islam. **Exports:** Oil, dates, fruit, and tobacco. **Climate:** Extremely hot. ☐ The ancient seafaring nation of Oman (o mahn') occupies the southeastern corner of the Arabian Peninsula. A detached tip of Oman juts into the Strait of Hormuz, separating the Persian Gulf and the Gulf of Oman. Oman's population and agriculture are concentrated on the northern and southern coasts where there is fertile land and moderate rainfall. A network of ancient underground canals provides additional water. Oman is rapidly modernizing.

## QATAR ☆ F

**Area:** 4,250 sq. mi. (11,008 km²). **Population:** 350,000. **Capital:** Doha, 195,000. **Government:** Emirate. **Language:** Arabic. **Religion:** Islam. **Exports:** Oil. **Climate:** Very hot and dry. ☐ Qatar is a barren, oil-rich peninsula that juts from its boundary with Saudi Arabia into the Persian Gulf. Oil has transformed Qatar from a poor, sea-dependent country to a bustling, almost entirely urban society that is even richer, per capita, than Kuwait. Less than a third of the rapidly expanding population is native-born. The ruling emir is a member of a family that has been in power for over 100 years.

## SAUDI ARABIA ☆ G

**Area:** 830,000 sq. mi. (2,149,700 km²). **Population:** 12,600,000. **Capital:** Riyadh, 670,000. **Government:** Monarchy. **Language:** Arabic. **Religion:** Islam. **Exports:** Oil. **Climate:** Hot except for highlands. ☐ Saudi (saw dee' or sow dee') Arabia, which occupies most of the Arabian Peninsula, is the

world's top oil exporter (the US and USSR are the largest producers). The nation has about a quarter of the world's reserves. The government has used its enormous income to invest widely abroad and to build large industrial complexes at home. Agriculture in this hot, riverless country is limited to highlands bordering the Red Sea and to fertile desert oases. Nearly half the people are nomadic Bedouins, but no one lives in the southern Rub al Khali Desert—the "Empty Quarter." The presence of the cities of Mecca and Medina, Islam's holiest shrines, make Saudi Arabia the preeminent nation in the Arab world. Muhammed (Mohammed), the founder of Islam, was born in Mecca in 570 AD. The revelations he received from God (Allah), which cover all aspects of daily life, are written in the Koran. Muhammed believed that he was the last in the Judaic and Christian tradition of great prophets. Islam preaches the existence of a single God but tolerates other faiths. Five times a day, nearly a billion Muslims kneel in the direction of Mecca to pray. Many visit the city to pray at the Great Mosque and its most sacred shrine, the Kaaba, a small black cube-shaped building believed to have come from God. Because the Koran requires Muslims to visit Mecca at least once in their lives, over a million pilgrims arrive each year. Providing accommodations for that many people is an industry in itself. Medina, the city Muhammed governed and in which he died, is almost as holy as Mecca. Most Arab Muslims are members of the Sunni sect. Shiite Muslims make up half of the Iraqi and almost all of the Iranian populations; they are often a poorer and persecuted minority in the Arab countries. Saudi Arabia's moral qualifications to be the foremost Islamic nation have been challenged by militant Shiites in Iran because of its moderate stance toward the West. The two sects have an ancient disagreement over who inherited the leadership of Islam after the death of Muhammed. The Islamic religion bans idolatry and the depiction of living things in art. Representing humans, animals, or even plants realistically is thought to usurp the role of Allah as the sole creator. These restrictions have fostered the creation of remarkably decorative and colorful abstract motifs in Islamic art and architecture. Outstanding examples are found in the mosque (house of worship); its minarets (tall, slender towers), and the familiar Persian rug.

## UNITED ARAB EMIRATES ☆ H

**Area:** 32,285 sq. mi. (83,618 km²). **Population:** 1,400,000. **Capital:** Abu Dhabi, 250,000. **Government:** Emirate Federation. **Language:** Arabic. **Religion:** Islam. **Exports:** Oil and oil products. **Climate:** Very hot and dry. ☐ In 1971, Britain withdrew as the protector and foreign advisor of the Trucial States; seven independent emirates banded together to form a federation known as the United Arab Emirates, each with its own ruling emir. The largest emirate, and the wealthiest state in the world, is Abu Dhabi. The federation takes care of matters of defense, foreign policy, and economic planning.

## YEMEN ☆ I

**Area:** 207,050 sq. mi. (528,489 km²). **Population:** 9,800,000. **Government:** Republic. **Capital:** Sana, 280,000. **Language:** Arabic. **Religion:** Islam. **Exports:** Coffee, cotton, dates, tobacco, fish, and oil. **Climate:** Hot and humid. ☐ For over 300 years, Yemen (North Yemen) and South Yemen were ruled separately. In recent times, South Yemen (The People's Democratic Republic of Yemen) was the only communist country in the Arab world. Relations between the two Yemens were hostile. But when democratic change, prompted by events in eastern Europe, came to South Yemen, the nations agreed to merge in 1990. Sana is the political capital, and Aden, with its oil refineries and shipbuilding facilities, is the economic capital. Northern Yemen has tall mountains and fertile fields—an unusually green region on the Arabian Peninsula. Its importance in trade between Africa and Asia dates back to Biblical times. The best known export is coffee, named after the port of Al Mukha (formerly called Mocha). In the north, rival Muslim factions have disagreed over the country's development. Shiites, who have the political and religious power, have opposed the modernization policies of the Sunni middle class.

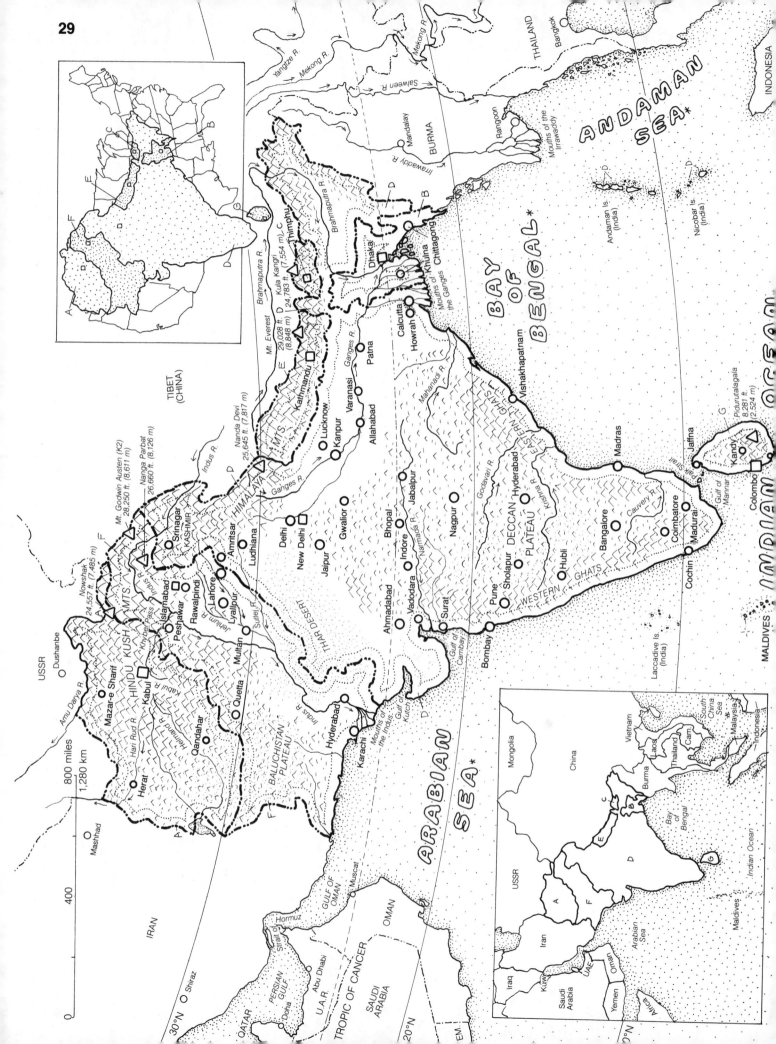

# ASIA: SOUTH & AFGHANISTAN

The Himalaya Mountains have been a natural barrier between Mongoloids speaking the Tibeto-Burman languages of Central Asia and Caucasoids speaking the Indo-Aryan languages of South Asia. Invasions and occupations have added many languages and dialects to South Asia's native tongues. English is the unifying language spoken in politics, business, education, and science. India, Bangladesh, and Pakistan formerly made up most of the British-Indian Empire. In 1947, predominantly Hindu India was granted independence. Pakistan was created as a homeland for Muslim minorities. East Pakistan seceded in 1971 to become Bangladesh. Afghanistan and the Maldives are also Islamic nations. Nepal, like India, is Hindu; Bhutan and Sri Lanka have Buddhist majorities. South Asia is the most densely populated area in the world. The huge populations are confined to coastal areas, river deltas, and the basins of the Ganges and Brahmaputra Rivers in India and Bangladesh and the Indus River in Pakistan. These countries have agricultural economies that depend upon the critical timing of summer monsoon rains. Local populations continue to grow rapidly, primarily because of a declining death rate due to improved health care and sanitation. There is a significant loss of life from animal attacks. Tigers, leopards, elephants, rhinoceroses, and poisonous snakes are just a few of a vast variety of wild and dangerous animals that fill the jungles of the subcontinent. An "abominable snowman" is reputed to be alive in the Himalayas, but its existence has never been proved.

## AFGHANISTAN A

**Area:** 253,000 sq. mi. (655,270 km²). **Population:** 17,000,000. **Capital:** Kabul, 925,000. **Government:** One-party republic. **Language:** Pashto; Dari. **Religion:** Islam. Exports: Natural gas, hides, dried fruits, and cotton. **Climate:** Dry; with continental extremes. ☐ Landlocked Afghanistan is a starkly beautiful mountainous nation. The only fertile land is found north of the Hindu Kush Mountains that cross the heart of the country. South of these towering peaks is a barren desert. For most of the 1980s the nation was gripped by a bloody civil war in which the Soviet Union intervened on behalf of the communist government in power. The rebel faction was made up of many different Muslim groups fighting to preserve a feudal way of life. These "Mujahadeen" (holy warriors), armed by the US, fought the Soviets to a standstill. Such religious-based resistance to centralized authority has been a common thread throughout Afghanistan's history. By the time the Soviets pulled out in 1989, well over a quarter of the civilian population had fled to Pakistan and Iran. For centuries, Afghanistan was considered the gateway to the riches of the Indian subcontinent and was the customary route for invading armies. The famed Khyber Pass across the Hindu Kush is still the major road for Afghan exports traveling to the Pakistani port of Karachi.

## BANGLADESH B

**Area:** 55,575 sq. mi. (143,940 km²). **Population:** 115,000,000. **Capital:** Dhaka, 2,500,000. **Government:** Republic. **Language:** Bengali. **Religion:** Islamic 85%; Hindu. Exports: Jute, tea, fish products, and hides. **Climate:** Tropical with very heavy rainfall. ☐ Bangladesh is essentially a huge delta (the world's largest), formed by five rivers, including the Ganges and Brahmaputra. The boat-filled countryside is subject to annual flooding brought on by some of the heaviest rainfall on the planet. Cyclone-driven tidal waves are an added threat; one such wave killed over a quarter of a million people in 1970. During the British occupation of India, Bangladesh was the Bengali-speaking, Muslim eastern half of the state of Bengal. It became the eastern portion of the newly formed Pakistan in 1947. Except for the Islamic religion, the two halves of the country, 1,000 mi. (1,600 km) apart, had nothing in common. After years of exploitation by the government in the West, East Pakistan seceded in 1971. The West attacked, but with the aid of India, Bangladesh ("Bengal Nation") was created. The nation's agricultural output is no match for the burgeoning population; this is one of the world's poorest and crowded areas. Bangladesh is the world's top producer of jute fibers used to make rope and sack-cloth.

## BHUTAN C

**Area:** 18,145 sq. mi. (46,980 km²). **Population:** 1,600,000. **Capital:** Thimphu, 24,000. **Government:** Monarchy. **Language:** Dzongkha (Tibetan dialect), Nepali. **Religion:** Buddhist 70%; Hinduism. Exports: Timber, fruit, and whiskey. **Climate:** Very wet; temperature varies according to altitude. ☐ This remote Himalayan kingdom wedged between India and Tibet is the closest thing to the fabled Shangri-la. Bhutan's (boo tahn') name means "land of the dragons," and the mythical animal graces the country's flag. Most of the population lives in the foothills and river valleys. About two-thirds of the people are Buddhists of Tibetan ancestry. Many monks live in hundreds of fortress-like monasteries. Until the 1960s, Bhutan was an isolated, almost totally illiterate nation. A program of modernization is now in progress. Road and air travel to India has been improved. India serves as Bhutan's protector.

## INDIA D

**Area:** 1,270,000 sq. mi. (3,289,300 km²). **Population:** 825,000,000. **Capital:** New Delhi, 4,000,000. **Government:** Republic. **Language:** Hindi and English (official); 850 other languages and dialects. **Religion:** Hinduism 83%, Muslim 11%. Exports: Iron ore, tea, cotton, hides, textiles, and rubber. **Climate:** Tropical with three seasons: cool, hot, and wet. ☐ India, the world's largest democracy, is a third the size of the US but has more people than any country except China. Paradoxically, this poor, predominantly agricultural country is rapidly becoming a major industrial nation and is one of the leaders in producing scientists and skilled technicians. India is unified by the Hindu religion but divided by over 800 languages and dialects. The Hindu religion, one of the world's oldest, supports a caste system with a rigid class structure that determines how the members of each caste shall live. One can never leave the caste of birth. There are four main castes, each with hundreds of subcastes. A person's standing in society depends upon his or her caste. Fifteen percent of Hindus are "untouchables": their unfortunate position is below the entire caste system. Modern laws prevent discrimination based on caste, but age-old traditions die slowly and class distinction remains a way of life. All Hindus believe in reincarnation in which it is possible for a human to return as an animal in the next life. Therefore, most Hindus do not eat meat. Cows are considered sacred and are allowed to wander through the business districts of major cities and to graze on valuable farmland. Other religions are represented in India: Muslims (11%) live mainly in the north; Christians (3%) live in the northeast; bearded, turban-wearing Sikhs (2%) have violently demanded greater autonomy in the northern state of Punjab; Buddhists (1%) were once in the majority; and Jains (1%) extend the reverence for life to all living creatures. Southern Indians are dark-skinned descendents of the Dravidians, the earliest known inhabitants of India. They were driven south by the Aryans, the ancestors of the light-skinned northern Indians. They represent completely different cultures.

India is bordered on the north by the Himalayas. The fertile region just to the south is the world's largest alluvial plain. This densely populated region consists of three river basins: the Indus, the Brahmaputra, and the temple-lined Ganges. Pilgrims come to bathe and spiritually cleanse themselves in the sacred waters of the Ganges. The triangular Indian Peninsula is a tropical plateau (The Deccan Plateau) rimmed by mountain ranges called the Western and Eastern Ghats. Ten cities in India have over 1 million residents. Many have modern sections, built by the British, that are currently occupied by wealthy or politically influential Indians. Thousands of homeless people bed down on the streets of the major cities. At Agra, in northern India, stands the Taj Mahal, one of the world's most beautiful structures. The white marble building was built as an Islamic tomb for a Indian prince and his wife. The Indian nonviolent movement for independence from Great Britain, led by Mohandas Gandhi, set an example for the American civil rights activists of the 1960s. But India has not been nonviolent in boundary disputes with China and Pakistan. After many battles between India and Pakistan over the beautiful state of Kashmir, a United Nations-mediated treaty established the current boundary lines. In 1990, tensions flared again as India accused Pakistan of aiding the Muslim separatist movement in Kashmir.

**MALDIVES.** This Islamic nation, consisting of 2,000 coral atolls, lies about 300 mi. (480 km) southwest of the tip of India. Most of the 200,000 people, who inhabit 200 of the islands, are descendents of Sri Lankans. Male (35,000) is the capital. Fish, coconuts, and tourism are the main industries.

## NEPAL E

**Area:** 54,588 sq. mi. (141,383 km²). **Population:** 18,500,000. **Capital:** Kathmandu, 210,000. **Government:** Monarchy. **Language:** Nepali 50%; many others. **Religion:** Hinduism 90%, Buddhism 10%. Exports: Food products, timber, and hides. **Climate:** Varies from alpine to tropical. ☐ The Himalayas (which include eight of the world's ten tallest peaks) occupy 90% of Nepal (nuh pawl'). The terrain consists mostly of mountain slopes. The country is less than 100 mi. (160 km) wide. It drops from snowy Himalayan peaks to a swampy tropical plain on the southern border. Nepal is the birthplace of Gautama Buddha (560 BC), the founder of Buddhism. The famous Sherpa guides accompany many mountain-climbing expeditions originating in Nepal. The country's renowned Gurkha soldiers have distinguished themselves in the British and Indian armies—Nepal is the only state in the region to have successfully resisted British occupation. Democracy and the hope of prosperity are coming to this very poor, mountain kingdom.

## PAKISTAN F

**Area:** 310,400 sq. mi. (803,936 km²). **Population:** 115,000,000. **Capital:** Islamabad, 225,000. **Government:** Republic. **Language:** Urdu. **Religion:** Islam. Exports: Natural gas, cotton products, textiles, carpets, and rice. **Climate:** Very dry and continental. ☐ In a land where monsoon winds blow hot or cold but almost never wet, Pakistan is completely dependent on the mighty Indus River and its six major tributaries. The rivers provide water for the world's largest irrigation system. The Indus Valley was the site of many advanced ancient South Asian civilizations. The largest industry is cotton and cotton goods. Most exports pass through the former capital, Karachi (5,250,000), on the Arabian Sea. Pakistan is a nation of many ethnic groups, each with its own language; fewer than 10% of the people speak Urdu, the official tongue. The nation was created as a homeland for Muslim minorities living in British India. Though a conservative Islamic nation, where the public presence of women is severely restricted, Pakistan elected Benazir Bhutto, the daughter of a former leader, as Prime Minister in 1988. She was removed from office in 1990 amid charges of corruption and incompetence.

## SRI LANKA G

**Area:** 25,330 sq. mi. (65,605 km²). **Population:** 18,500,000. **Capital:** Colombo, 600,000. **Government:** Republic. **Language:** Sinhala; Tamil. **Religion:** Buddhist 75%; Hinduism 18%. Exports: Tea, rubber, coconuts, and graphite. **Climate:** Tropical. ☐ Sri Lanka (sree lahn' kuh), formerly Ceylon, is a beautiful tropical island linked to the Indian mainland by "Adam's Bridge," a 20 mi (32 km) chain of sandy islands. The Buddhist Sinhalese majority (75%) and the Hindu Tamil minority (18%) originally came from India. Several Tamil groups have been waging a guerrilla war for an independent state in the north. The factional war has destroyed Sri Lanka's earlier promise of becoming an economically prosperous nation. Most people live in the wet and hilly southwest region which is ideal for growing tea—Sri Lanka is the world's number two producer and the leading producer of high-quality graphite.

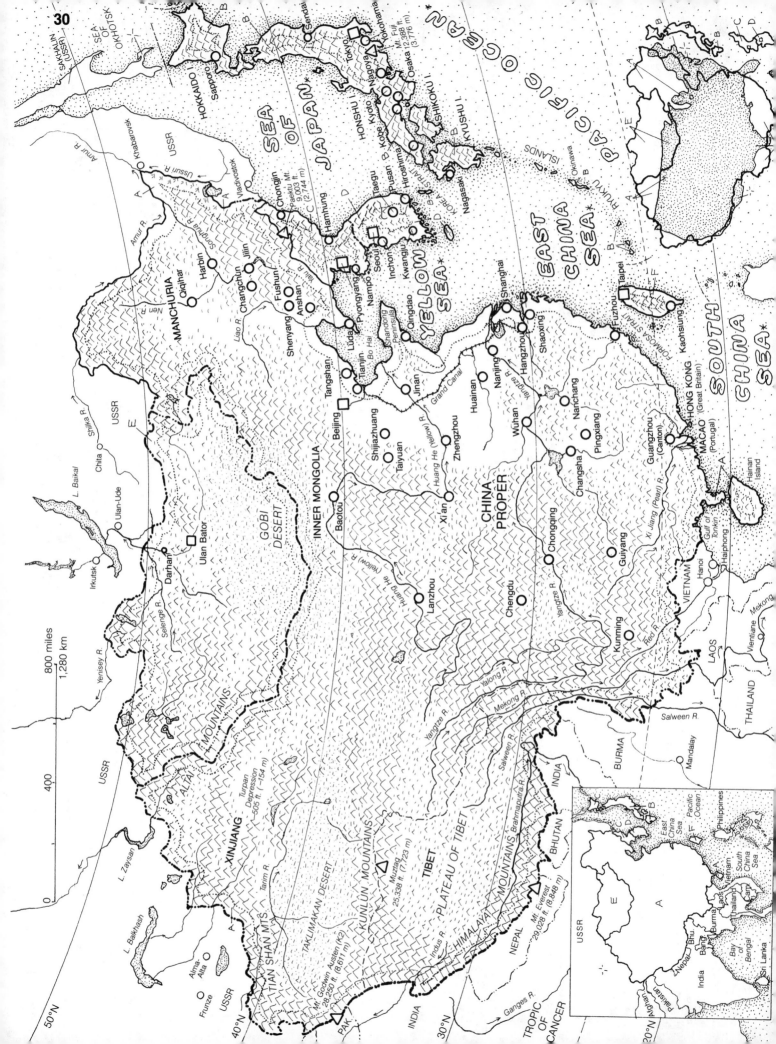

# 30

SEA OF OKHOTSK (USSR)

SAKHALIN (USSR)

SEA OF JAPAN

PACIFIC OCEAN

HOKKAIDO

Sapporo

Sendai

Tokyo

Yokohama

Nagoya

Mt. Fuji (12,388 m)

Osaka (3,776 m)

Kyoto

Kobe

HONSHU

SHIKOKU I.

Hiroshima

Nagasaki

KYUSHU I.

RYUKYU ISLANDS

Okinawa

EAST CHINA SEA*

Khabarovsk

Amur R.

Ussuri R.

USSR

Vladivostok

Chongjin

Paektu Mt. 9,003 ft (2,744 m)

Hamhung

KOREA

Pusan

Taegu

Kwangju

KOREA STRAIT

Amur R.

Nen R.

Songhua R.

MANCHURIA

Qiqihar

Harbin

Jilin

Changchun

Fushun

Anshan

Shenyang

Liao R.

Lüda

Tianjin

Bo Hai

Pyongyang

Nampo

Seoul

Inchon

Shandong Peninsula

Qingdao

YELLOW SEA*

Shanghai

Shaoxing

Hangzhou

Fuzhou

FORMOSA STRAIT

Taipei

Kaohsiung

SOUTH CHINA SEA*

Shilka R.

Chita

USSR

Ulan-Ude

L. Baikal

Irkutsk

Yenisey R.

Selenge R.

Darhan

Ulan Bator

GOBI DESERT

INNER MONGOLIA

Baotou

Beijing

Tangshan

Shijiazhuang

Taiyuan

Jinan

Grand Canal

Huang He (Yellow) R.

Zhengzhou

Huainan

Nanjing

Wuhan

Nanchang

Pingxiang

Changsha

Xi'an

CHINA PROPER

Chongqing

Yangtze R.

Guiyang

Guangzhou (Canton)

Xi Jiang (Pearl) R.

HONG KONG (Great Britain)

MACAO (Portugal)

Hainan Island

Gulf of Tonkin

Haiphong

VIETNAM

Hanoi

Red R.

LAOS

Vientiane

THAILAND

Mekong R.

Kunming

Chengdu

Lanzhou

Huang He (Yellow) R.

800 miles

1,280 km

400

L. Balkhash

L. Zaysan

USSR

Alma-Ata

Frunze

XINJIANG

TIAN SHAN MTS.

ALTAI MOUNTAINS

Tarim R.

Turpan Depression –505 ft (–154 m)

TAKLIMAKAN DESERT

Muztag 25,388 ft (7,723 m)

KUNLUN MOUNTAINS

Mt. Godwin Austen (K2) 28,250 ft (8,611 m)

PLATEAU OF TIBET

TIBET

Indus R.

PAK.

INDIA

NEPAL

Mt. Everest 29,028 ft (8,848 m)

HIMALAYA MOUNTAINS

Brahmaputra R.

Salween R.

BHUTAN

INDIA

Ganges R.

BURMA

Mandalay

Salween R.

Yalong R.

Mekong R.

TROPIC OF CANCER

50°N

40°N

30°N

20°N

A B C D E F

Pacific Ocean

East China Sea

South China Sea

Philippines

USSR

Bay of Bengal

India

Nepal

Bhu.

Bang.

Burma

Thailand

Laos

Camb.

Vietnam

Afghan.

Pakistan

Sri Lanka

# ASIA: FAR EAST

Included among the extremely crowded nations of the Far East is the world's most deserted country, Mongolia. The Mongolians were the only Asian communists to be influenced by the European reform movements of the 1980s. Though still politically hard-line, communist China has made economic changes patterned after the successful Asian free-market nations: Japan and the "Little Dragons" (Hong Kong, South Korea, and Taiwan). Those booming economies have had to move factories to China and other Asian nations in search of larger labor pools. Ancient China dominated this region. Surrounding nations still show a Chinese influence on their language, religion, and culture. The early 20th century marked the beginning of Japanese expansionism. During World War II, Japan's empire began at the Indian border and covered Manchuria (which Japan called Manchukuo); eastern China; Korea (Chosen); Taiwan (Formosa); the Philippines; countries of southeast Asia; Indonesia; the islands of the western Pacific; and the outer islands of Alaska.

## CHINA A

**Area:** 3,685,000 sq. mi. (9,544,150 km²). **Population:** 1,115,000,000. **Capital:** Beijing (Peking), 6,000,000. **Government**: Republic (communist). **Language:** Mandarin (Northern Chinese); Cantonese. **Religion:** Confucianism 20%; Buddhism 7%; Taoism 2%. **Exports:** Manufactured goods, cotton, silk, textiles, tea, oil, and asbestos. **Climate:** The north and west are dry and continental; the east is moderate; the southeast is subtropical, with heavy summer rain. ☐ China, the world's oldest civilization, is the third largest nation (after the USSR and Canada) and has the largest population. A billion people are crowded into the eastern 15% of China—one out of every five human beings on the planet. Though 80% of the population live in farms and villages, more than 30 cities have over a million residents. China is struggling to curb its population growth by severely penalizing families who have more than one child. There are many minority groups in China, but their total number, though in excess of 50 million, represents less than 6% of a primarily Han Chinese population. Over two-thirds of the people speak the northern dialect foreigners call Mandarin. All Chinese write with the same pictorial characters. A western alphabet that spells Chinese words phonetically is now used for foreign communications: Beijing, not Peking, is the capital, and Mao Zedong, not Mao Tse-tung, was the leader of the Communist revolution. China is composed of 22 provinces and 5 autonomous regions. Tibetan refugees, who have witnessed the deliberate obliteration of their culture by the Chinese government, would not agree that their homeland on the southwestern border is "autonomous." The bleak Tibetan plateau is the world's highest inhabited region (15,000 ft., 4,573 m). Its southern border is rimmed by the Himalayas and Mt. Everest, the world's tallest peak. The *northeast,* formerly called Manchuria, is a major industrial and agricultural region. The *north-central uplands,* called Inner Mongolia, includes part of the Gobi Desert. The *northwest,* a sparsely populated area called Xinjiang, includes the Tian Shan Mountains, the Taklimakan Desert, and the Turpan Depression. *China proper* consists of the eastern lowlands and the central and southeastern uplands. The humid subtropical climate of south China permits multiple rice harvests. The fertile valleys of eastern China have been created by the Huang He and Yangtze Rivers, which originate in Tibet. The Huang He is called the Yellow River because its waters are colored by the yellow soil it transports. Accumulations of silt can raise the riverbed, causing devastating floods and even a change of course. The Yangtze flows through the heart of China's trade and industry and reaches the sea at one of the world's busiest ports: Shanghai. Virtually every village and city in southeast China is located on one of the hundreds of rivers and canals that serve as local highways. For centuries, junks with quilted sails and sampans resembling seagoing quonset huts have been used for fishing, transportation, and housing. Cormorants are still used for fishing: nooses around the necks of these sea-diving birds prevent them from swallowing their catch. Close to the east coast is the longest artificial waterway in the world, the Grand Canal (1,105 mi., 1,768 km), which links Beijing to Hangzhou. The world's longest structure is the Great Wall which snakes across the northern mountains and valleys for 1,500 mi. (2,400 km). It was designed to keep out marauding nomads and was built over a period of 2,000 years, ending in the 17th century. Parts of the wall are 30 ft. (11.6 m) wide and 50 ft. (15 m) high. The communist revolution of 1949 eliminated disease and starvation but failed to raise the standard of living. In the 1980s, China's leaders allowed the introduction of economic reforms and free market incentives, but the brutal suppression of demands for democracy in 1989 has placed in jeopardy future foreign investment.

**HONG KONG.** This "Little Dragon" does business around the clock. Six million people (98% Chinese) are packed into this tiny peninsula and island group (410 sq.mi., 1,062 km²) off the south China coast. Despite assurances by the Chinese government, residents of this British Crown Colony are apprehensive about their future: ownership will revert to China in 1997. Rising wages and a labor shortage have forced Hong Kong to transfer most of its production to the Chinese mainland. Hong Kong still functions as one of the Far East's centers for trade, finance, and tourism. Tall modern buildings overlook the busy harbor separating the island's capital, Victoria (1,000,000), from Kowloon, its companion city on the peninsula.

**MACAO.** On the south China coast, 40 mi. (64 km) west of Hong Kong, is this Portuguese overseas province that, in 1999, will also revert to China. Few Portuguese live here, but their influence is evident in the city's Mediterranean appearance. Tourists jetfoil from Hong Kong to gamble in Macao's casinos. In the past, when it was a center for drug smuggling and foreign intrigue, Macao was called the "Casablanca of the Far East."

## JAPAN B

**Area:** 145,745 sq. mi. (377,480 km²). **Population:** 122,000,000. **Capital:** Tokyo, 8,500,000. **Government:** Constitutional monarchy. **Language:** Japanese. **Religion:** Shintoism; Buddhism. **Exports:** Cars, trucks, ships, consumer electronics, chemicals, fish, and textiles. **Climate:** Temperate, with heavy seasonal rainfall; north is cold. ☐ The "Land of the Rising Sun" has risen from the ruins of World War II to become the world's second largest industrial economy. The military-dominated monarchy which directed Japan's war machine was replaced by US-imposed democratic institutions as well as aid to rebuild its factories. Despite having to import all of its raw materials, revenues from product sales have made Japan the world's number one financial power and foreign investor. Living space is at a premium—half the population lives on 2% of the land. Three quarters of the population live on Honshu Island; the metropolitan area of Tokyo and Yokohama has over 25 million people. The fourth and northernmost island, Hokkaido, is a winter recreation area and the home of Japan's only minority group, the Ainu tribe. These 15,000 Caucasians are descendants of the country's original inhabitants. Hokkaido is cooled by the offshore Oyashio Current; the southern islands are warmed by the Kuroshio (Japan) Current (Plate 44). Every square inch of arable land is carefully cultivated. The Shinto religion reveres many gods in nature. The reverence for natural forms is readily seen in Japanese architecture and landscaping. Most people practice a blend of religions and observe the holidays of both Buddhism and Shintoism. There is also a strong Confucian influence from China which teaches respect for family and authority. Today's Japanese culture is a fascinating combination of the traditional and the modern. This can be seen in the summer ritual of climbing Mt. Fuji, the sacred extinct volcano. What was formerly the practice of religious pilgrims is now a recreational pursuit of tens of thousands who seek the exercise and the magnificent view. Cone-shaped, snowcapped Fuji is probably the most commonly reproduced graphic symbol of Japan.

When the defeated Japanese gave up "Chosen" in 1945, the North was occupied by the Soviets and the South by the Americans. When these forces withdrew, the peninsula remained permanently divided. In 1950, communist North Korea invaded South Korea. US troops, aided by United Nations forces, came to the aid of the South. After three years of fighting, the armies were stalemated along the 38th parallel, the original dividing line. For nearly 40 years, American troops have protected South Korea. The North and the South are about the same size, but the North has mineral wealth and South Korea has arable land and twice the population. The Korean language shows the influence of earlier Chinese and Japanese occupations.

## NORTH KOREA C

**Area: 46,700 sq. mi. (120,953 km²). Population:** 22,000,000. **Capital:** Pyongyang, 2,750,000. **Government:** Republic (communist). **Language:** Korean. **Religion:** Officially discouraged. **Exports:** Fish, graphite, iron ore, copper, lead, and zinc. **Climate:** Very cold winters. ☐ Under Kim Il Sung's 45-year reign, North Korea has been the closest thing to George Orwell's *1984.* Sung's picture is everywhere; there is daily indoctrination; and radios can tune only to the government's station. The nation has a large industrial base, plenty of hydroelectric power, and trades in the communist world.

## SOUTH KOREA D

**Area:** 38,025 sq.mi.(98,485 km). **Population:** 44,000,000. **Capital:** Seoul, 8,400,000. **Government:** Republic. **Language:** Korean. **Religion:** Buddhism; Confucianism; Christianity. **Exports:** Consumer electronics, cars, textiles, and chemicals. **Climate:** Moderate, with hot, damp summers. ☐ In two decades, South Korea has changed from a poor, agricultural nation to one of the "Little Dragons." The accumulation of great wealth is being flaunted, contrary to Confucian tradition (which relegates the merchant to the lowest class), and is fueling tensions between rich and poor. The Christian minority (25%) is the largest on the Asian mainland. The Unification Church, which sponsors the "Moonies" cult, was founded in South Korea by Sun Myung Moon.

## MONGOLIA E

**Area:** 604,248 sq. mi. (1,565,002 km²). **Population:** 2,150,000. **Capital:** Ulan Bator, 350,000. **Government:** Republic. **Language:** Mongolian. **Religion:** Previously Buddhism. **Exports:** Livestock, animal products, fluorspar, and tungsten. **Climate:** Dry and extremely continental. Trees are even scarcer than people in the world's largest landlocked nation. The Gobi, the world's northernmost desert, occupies the southern third of the land. In the 13th century, a Mongolian, Ghengis Khan, reigned over the largest land empire in history, from eastern Europe to the coast of China. The barbarian empire dissolved, and the region became "Outer Mongolia" during the long period of Chinese control. Mongolia has been a staunch ally of the Soviet Union which helped liberate it from China in 1924. The once-nomadic people formerly herded livestock and lived in large, round, felt-covered tents called "yurts"; most now live in villages organized by the former communist government.

## TAIWAN F

**Area:** 13,910 sq. mi. (36,027 km²). **Population:** 21,500,000. **Capital:** Taipei, 2,500,000. **Government:** Republic. **Language:** Mandarin Chinese; dialects. **Religion:** Buddhism; Confucianism. **Exports:** Clothing, consumer electronics, and plastics. **Climate:** Subtropical. ☐ In 1949, Taiwan (formerly Formosa), became "Nationalist China" when Chiang Kai-shek and his army and followers fled to the island after the communist victory. For many years the Nationalists vowed to return to the mainland, 100 mi. (160 km) away. But in 1971, Communist China replaced Taiwan as the United Nation's representative of the Chinese people. Taiwan has since become one of the Far East's extremely prosperous "Little Dragons." Most residents live and work in the coastal areas. Mountain forests provide over half the world's supply of camphor, a substance used in chemicals, pharmaceuticals, and cosmetics.

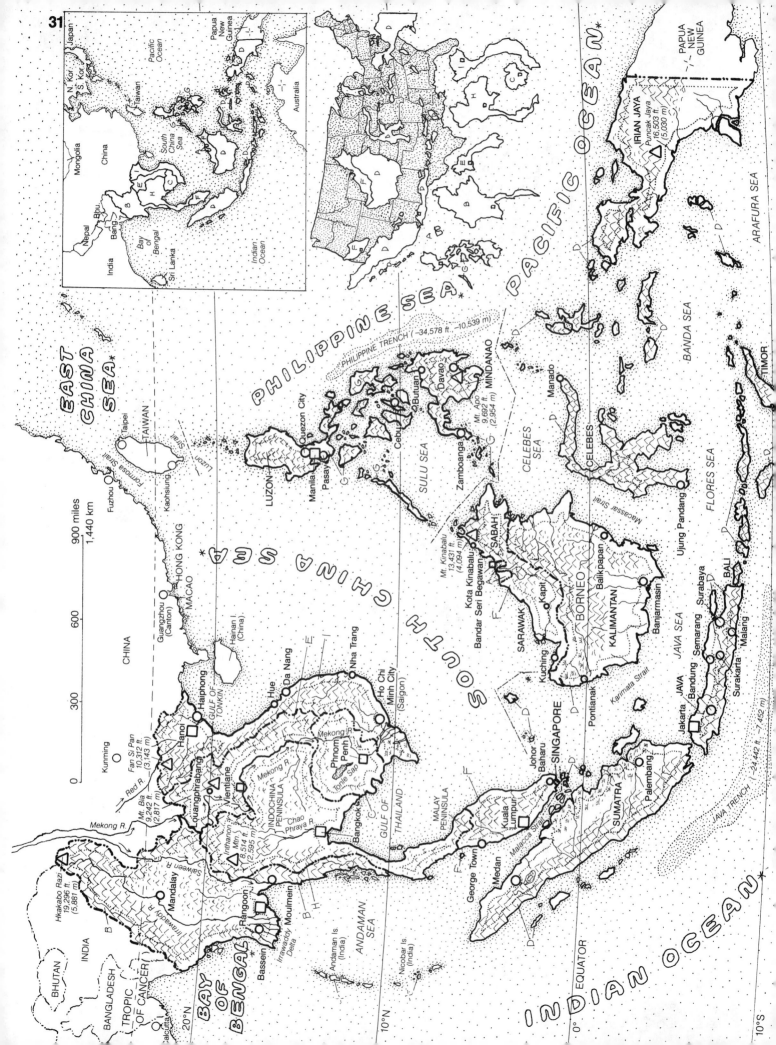

31

PACIFIC OCEAN*

PHILIPPINE SEA*

EAST CHINA SEA*

SOUTH CHINA SEA*

ARAFURA SEA

BANDA SEA

FLORES SEA

CELEBES SEA

SULU SEA

GULF OF TONKIN

GULF OF THAILAND

BAY OF BENGAL

ANDAMAN SEA

JAVA SEA

INDIAN OCEAN*

JAVA TRENCH (−24,442 ft. −7,452 m)

PHILIPPINE TRENCH (−34,578 ft. −10,539 m)

PAPUA NEW GUINEA

IRIAN JAYA
Puncak Jaya
16,503 ft
(5,030 m)

MINDANAO
Mt. Apo
9,692 ft.
(2,954 m)
Davao
Butuan
Zamboanga
Cebu

CELEBES
Manado

Manila
Pasay
Quezon City
LUZON

Taipei
TAIWAN
Kaohsiung
Formosa Strait
Luzon Strait

Fuzhou

Guangzhou (Canton)
HONG KONG
MACAO
Hainan I. (China)

CHINA
Kunming

SABAH
Mt. Kinabalu
13,431 ft.
(4,094 m)
Kota Kinabalu
Bandar Seri Begawan

SARAWAK
Kapit
Kuching

BORNEO
KALIMANTAN
Balikpapan
Banjarmasin
Pontianak

Karimata Strait

Makassar Strait
Ujung Pandang
BALI
Surabaya
Malang
Surakarta
Semarang
Bandung
JAVA
Jakarta

Palembang
SUMATRA

SINGAPORE
Johor Baharu
Kuala Lumpur
MALAY PENINSULA
George Town
Medan

Malacca Strait

TIMOR

Haiphong
Hanoi
Da Nang
Hue
Nha Trang
Ho Chi Minh City (Saigon)
Phnom Penh
Tonle Sap
Mekong R.
INDOCHINA PENINSULA
Chao Phraya R.
Bangkok

Vientiane
Louangphrabang
Fan Si Pan
10,312 ft.
(3,143 m)
Mt. Bia
9,242 ft.
(2,817 m)
Red R.
Mekong R.

Inthanon Mtn.
8,514 ft.
(2,595 m)

Mandalay
Salween R.
Irrawaddy R.
Rangoon
Moulmein
Bassein
Irrawaddy Delta

Hkakabo Razi
19,296 ft.
(5,881 m)

Andaman Is. (India)
Nicobar Is. (India)

BHUTAN
INDIA
BANGLADESH
Calcutta

TROPIC OF CANCER

20°N

10°N

EQUATOR

10°S

900 miles
1,440 km
600
300
0

Mongolia
China
Nepal
Bhu.
Bang.
India
Sri Lanka
Bay of Bengal
Indian Ocean
South China Sea
Pacific Ocean
Japan
N. Kor.
S. Kor.
Taiwan
Papua New Guinea
Australia

# ASIA: SOUTHEAST

The nations of Southeast Asia are located on the Indochina and Malay peninsulas, in the Malay Archipelago, and in the Philippines. The region is mountainous, forested, very warm, extremely humid, and seasonally drenched. Rivers and canals serve as roads in many areas. Native homes are perched on stilts for protection against flooding and wild animals. The watery landscape sustains the basic diet—fish and rice. Malaysia, Indonesia, and Thailand are the world's leading tin and rubber producers. Rubber trees were imported from South America in the 19th century. Tropical hardwood forests of teak, ebony, mahogany, and rosewood take 200 years to grow but are being cut down at an ever-increasing rate. Elephants are used as "trucks" by the timber industry, and water buffalos serve as local farm "tractors." Brunei, Malaysia, and Indonesia are Islamic nations. The countries on the Indochina peninsula are Buddhist; temples and monasteries can be seen everywhere, including communist Cambodia, Laos, and Vietnam (all formerly part of French Indochina). The British occupied Brunei, Burma, Malaysia, and Singapore. The Dutch governed Indonesia (the Dutch East Indies) for 300 years. Spain and the US governed the Philippines, the only Christian nation in eastern Asia. Thailand is the one country in the region to avoid colonization.

## BRUNEI A

**Area:** 2,226 sq. mi. (5,765 km²). **Population:** 230,000. **Capital:** Bandar Seri Begawan, 55,000. **Government:** Sultanate. **Language:** Malay; English. **Religion:** Islam. **Exports:** Oil,natural gas, and rubber. **Climate:** Hot and wet. ☐ Brunei is similar to an oil-rich Arab state: the ruling sultan is the world's richest man; wealthy residents are mostly Muslims; there are no taxes; and social services are free. The Chinese minority (20%) operate most of the businesses. Brunei is split in two by a strip of land belonging to Malaysia.

## BURMA B

**Area:** 261,750 sq. mi. (677,933 km²). **Population:** 42,000,000. **Capital:** Rangoon, 2,500,000. **Government:** Socialist republic. **Language:** Burmese. **Religion:** Buddhist. **Exports:** Rice, teak, sugar, precious gems, and rubber. **Climate:** Tropical. ☐ Burma has been politically isolated and economically undeveloped, partly due to the surrounding mountains but primarily because of its socialist government. The Irrawaddy River, the nation's main highway, flows through the agricultural and population centers. Its delta is the world's leading rice-producing area. At one of the mouths of the river is the capital, Rangoon. The city grew up around Shwe Dagon, a 2,500-year-old pagoda that is strikingly decorated with tall, ornately molded, gold leaf-covered spires. It is the most magnificent of Burma's many Buddhist temples. In the surrounding mountains live over 100 ethnically different native tribes. Burmese women in the urban areas are among the most emancipated in Asia.

## CAMBODIA C

**Area:** 69,890 sq. mi. (181,015 km²). **Population:** 7,000,000. **Capital:** Phnom Penh, 700,000. **Government:** Republic (communist). **Language:** Khmer. **Religion:** Buddhism. **Exports:** Rubber, rice, and timber. **Climate:** Tropical. ☐ Cambodia (also called Kampuchea) is saucer-shaped; its large central plain region surrounds the Tonle Sap (the "Great Lake"). The lake provides fish protein as well as flood water needed for rice farming. Most Cambodians are Khmers, the region's oldest ethnic group. From 700 to 1200 AD they dominated Indochina. In 1975, radical communists called the Khmer Rouge unleashed a reign of terror that killed over a million Cambodians. The "killing fields" were filled with the bodies of potential reactionaries: intellectuals, priests, reporters, professionals, teachers, librarians, merchants, and their families. Cities were emptied; those residents not killed were sent to work in the country. In 1979, Vietnam drove the Khmer Rouge out of power. But Vietnam withdrew in 1989; different factions (including the Khmer Rouge) are vying for power in an atmosphere of uncertainty.

## INDONESIA D

**Area:** 741,100 sq. mi. (1,919,449 km²). **Population:** 175,000,000. **Capital:** Jakarta, 6,000,000. **Government:** Republic. **Language:** Bahasa Indonesia. **Religion:** Islam. **Exports:** Oil, timber, rubber, tin, coffee, palm oil, and sugar. **Climate:** Tropical. ☐ Indonesia occupies more than 13,000 islands straddling the equator. The Malay Archipelago is the world's longest island chain (it is wider than the US). Indonesia has the world's fifth largest population, making it the largest Islamic nation. Nearly two-thirds of its people live on Java, an island the size of Alabama. A small Chinese minority (2%) run many of the nation's businesses. Buddhist and Hindu temples are reminders of a pre-Islamic past. Indonesians still believe in spirits and tend to follow their own brand of Islam. Bali, an island off the coast of Java, is a Hindu enclave filled with magnificent temples. Tourists are drawn by Bali's great beauty and its highly advanced culture based upon music and dance. Kalimantan, the largest state, occupies most of Borneo, a mineral-rich island that has made Indonesia the leading oil producer in eastern Asia. Borneo is the home of the Dyak tribes, formerly headhunters. Some of the Papuans living in Irian Jaya, which occupies the western half of New Guinea, are still living in the stone age. Indonesia experiences many earthquakes and has the most active volcanoes (nearly 100) of any country. In 1883, the greatest explosion in history destroyed most of the island of Krakatoa. The blast was heard for thousands of miles, and volcanic dust affected the earth's weather for several years.

## LAOS E

**Area:** 91,430 sq. mi. (236,804 km²). **Population:** 3,900,000. **Capital:** Vientiane, 185,000. **Government:** Republic (communist). **Language:** Lao. **Religion:** Buddhism. **Exports:** Tin, coffee, and timber. **Climate:** Tropical. ☐ Landlocked Laos (lah' os) is a poor nation wedged between two historically dominant neighbors, Thailand and Vietnam. The Mekong River, which defines most of the western border, is the main highway in a country of few roads and no railroads. Most of the diverse people live in the Mekong Valley. Laos was heavily bombed by the US in the Vietnam War when its "Ho Chi Minh Trail" was used by the North Vietnamese to supply their forces in South Vietnam. Thousands of refugees have left Laos since the communist takeover in 1975.

## MALAYSIA F

**Area:** 127,318 sq. mi. (329,754 km²). **Population:** 17,000,000. **Capital:** Kuala Lumpur, 550,000. **Government:** Constitutional monarchy. **Language:** Bahasa Malaysia; Chinese. **Religion:** Islam 50%; Buddhism 35%. **Exports:** Oil, rubber, tin, palm oil, and spices. **Climate:** Tropical. ☐ Malaysia (muh lay' zhuh) is a federation of independent sultanates. Two of them, oil-rich Sarawak and Sabah, occupy the northern part of Borneo. Eighty percent of the population lives on the southern end of the Malay Peninsula. Tensions have existed between the politically powerful Malayan Muslim majority and the economically strong Chinese Buddhists. The nation is the top producer of rubber, tin, and palm oil. Malaysia has a rapidly expanding industrial sector and is a likely candidate for Asia's prosperous club of "Little Dragons."

## PHILIPPINES G

**Area:** 115,835 sq. mi. (300,013 km²). **Population:** 61,000,000. **Capital:** Manila, 1,750,000. **Government:** Republic. **Language:** Pilipino and English. **Religion:** Roman Catholic. **Exports:** Timber, copra, abaca, sugar, and copper. **Climate:** Tropical and humid. ☐ The Philippines Archipelago consists of 7,000 mountainous, earthquake-prone, volcanically active islands. Only 10% are inhabited. The largest, Luzon and Mindanao, are home to two-thirds of the population. The people are mostly of Malay and some Polynesian ancestry. The Philippines resembles Latin American countries in many ways: it was a Spanish colony for over three centuries (Magellan discovered it in 1521); it is predominantly Catholic; the capital, Manila, is a contrast between modern architecture and frightful slums; a left-wing guerrilla movement has arisen in response to widespread poverty and the absence of land reform; and corrupt presidents have become rich while in office. The US acquired the Philippines from Spain after the Spanish-American War of 1898 and governed the colony for 48 years. An American-style educational system was introduced, and English is widely used. The Philippines is the leading producer of copra (coconut meat) and abaca (Manila hemp). Rice is the principal crop; some paddies in the mountains were terraced over 2,000 years ago.

**SINGAPORE.** Singapore has been the dominant shipping port of Southeast Asia since 1819, when an agent of the British East India Company created it for the purpose of trade. The tiny (227 sq.mi., 588 km) island republic is connected to the tip of the Malay Peninsula by a causeway less than a mile long. It is the richest of Asia's "Little Dragons." Singapore is a center for finance, general manufacturing, consumer electronics, oil and rubber processing, shipbuilding, and shipping. Three-fourths of the 2,700,000 people are Chinese, and almost all live in the city of Singapore.

## THAILAND H

**Area:** 198,450 sq. mi. (513,986 km²). **Population:** 56,000,000. **Capital:** Bangkok, 5,250,000. **Government:** Constitutional monarchy. **Language:** Thai. **Religion:** Buddhism. **Exports:** Rubber, tin, teak, bamboo, and handcrafts. **Climate:** Tropical. ☐ In Thailand (ty' land), formerly called Siam, almost everything happens on or close to water. The capital, Bangkok, is reminiscent of Venice; it is at the mouth of Chao Phraya River. The river basin, a fertile, alluvial plain, is the national "rice bowl" and the heart of a prosperous economy. Rivers are often filled with logs from Thailand's sizeable teak forests. Every village has its Buddhist temple, and each Thai boy is expected to spend a few months living the life of a monk. Thailand means "free country" and Thais have managed to enjoy 700 years of independence through skillful diplomacy. They gave military assistance to Britain and France in World War I in exchange for a guarantee of its sovereignty. Thailand is now burdened by refugees from its communist neighbors—Cambodia, Laos, and Vietnam.

## VIETNAM I

**Area:** 127,240 sq. mi. (329,551 km²). **Population:** 64,000,000. **Capital:** Hanoi, 2,500,000. **Government:** Republic (communist). **Language:** Vietnamese. **Religion:** Buddhism. **Exports:** Coal, timber, rubber, and rice. **Climate:** Tropical. ☐ Vietnam is mostly mountainous; its huge population is concentrated in two delta regions connected by a long, narrow strip of low coastal land. Hanoi, the capital, is located on the silt-colored Red River delta in the north. The Mekong River delta in the south lies 3,200 mi., (5,120 km) from its headwaters in Tibet. The nation's largest city is on the delta—French-flavored Ho Chi Minh City (3,500,000), formerly Saigon. Vietnamese communists under Ho Chi Minh led the resistance against the Japanese occupation in World War II. The French tried to reclaim their colony after the war, but they were defeated by the communists in 1954 after eight years of fighting. The country was then temporarily divided; elections were to determine reunification. The noncommunist South, backed by the US, refused to hold elections, fearing defeat by the immensely popular Ho Chi Minh. The Viet Cong guerrilla movement (Viet Cong) was formed to overthrow the unpopular South Vietnamese government, which was run by wealthy, pro-French Roman Catholics (Vietnam is 80% Buddhist). Vietnam harbors a traditional hatred of China because of 900 years of occupation; apparently unaware of this, the US mistook the communist-led nationalist movement for Chinese expansionism in Indochina and intervened massively to prevent the fall of South Vietnam. The North responded with its own intervention, and the Vietnam War raged on until the US withdrew in 1973. The South then collapsed, and Vietnam became the world's third largest communist country. Thousands of refugees, including the desperate "boat people," have left Vietnam since the war, many of them settling in the US. Vietnam and China remain enemies.

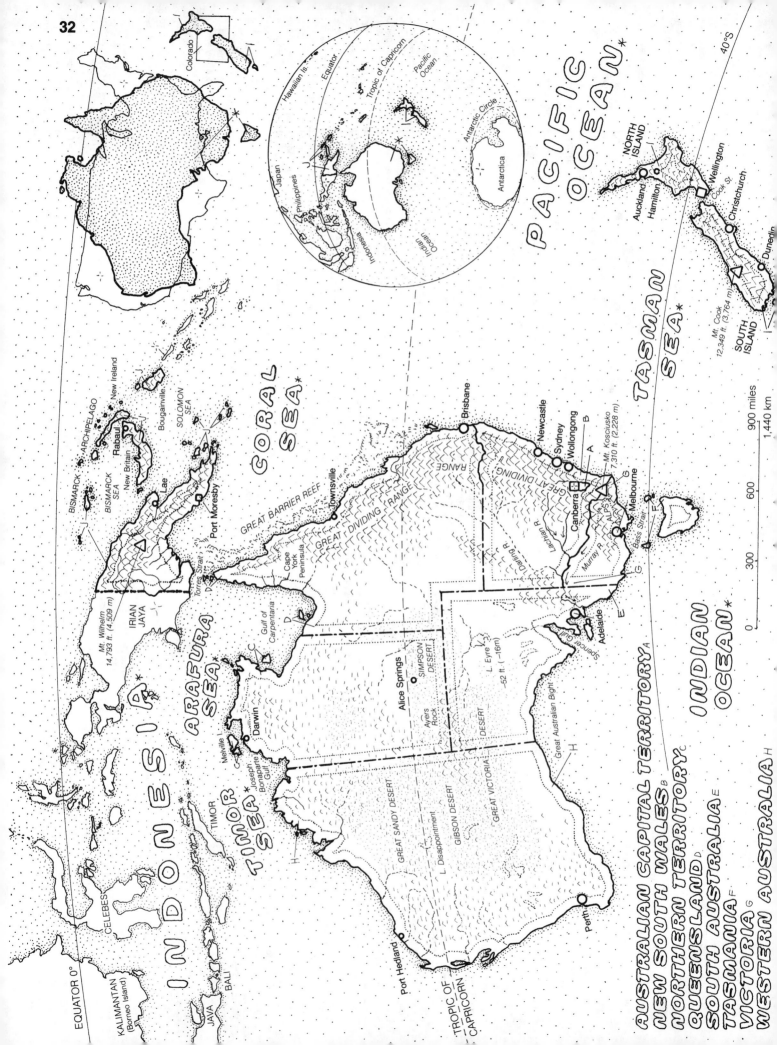

32

PACIFIC OCEAN*

Hawaiian Is.
Equator
Tropic of Capricorn
Pacific Ocean
Antarctic Circle
Japan
Philippines
Indonesia
Indian Ocean
Antarctica

Colorado

NORTH ISLAND
Auckland
Hamilton
Wellington
Cook St.
Christchurch
SOUTH ISLAND
Dunedin
Mt. Cook
12,349 ft. (3,764 m)

40°S

TASMAN SEA*

CORAL SEA*

Brisbane
Newcastle
Sydney
Wollongong
Mt. Kosciusko
7,310 ft. (2,228 m)

GREAT DIVIDING RANGE

GREAT BARRIER REEF
Townsville

Lachlan R.
Canberra
Darling R.
ALPS
Melbourne
Murray R.
Bass Strait

New Ireland
New Britain
Rabaul
BISMARCK ARCHIPELAGO
BISMARCK SEA
Lae
SOLOMON SEA
Bougainville
Port Moresby

Torres Strait
Cape York Peninsula
GREAT DIVIDING RANGE

Mt. Wilhelm
14,793 ft. (4,509 m)
IRIAN JAYA

INDONESIA*

ARAFURA SEA*

Gulf of Carpentaria

Alice Springs
Ayers Rock
SIMPSON DESERT

L. Eyre
-52 ft. (-16m)

Spencer Gulf
Adelaide
Great Australian Bight

Melville
Darwin
Joseph Bonaparte Gulf
TIMOR SEA*
TIMOR

CELEBES

EQUATOR 0°
KALIMANTAN (Borneo Island)

JAVA
BALI

GREAT SANDY DESERT
L. Disappointment
GIBSON DESERT
GREAT VICTORIA DESERT

TROPIC OF CAPRICORN

Port Hedland

Perth

INDIAN OCEAN*

AUSTRALIAN CAPITAL TERRITORYA
NEW SOUTH WALESB
NORTHERN TERRITORYC
QUEENSLANDD
SOUTH AUSTRALIAE
TASMANIAF
VICTORIAG
WESTERN AUSTRALIAH

900 miles
1,440 km
600
300
0

# OCEANIA I: AUSTRALIA, NEW ZEALAND, & PAPUA NEW GUINEA

A continent is surrounded by water; Oceania is water that surrounds a continent as well as thousands of tiny, widely scattered islands. Oceania, also called Australasia, encompasses a region 8,000 mi. (12,800 km) wide in the southern hemisphere, from Australia to the central Pacific Ocean. Australia is by far the largest of the thousands of islands. It is large enough to be considered a continent, but compared to other continents, it is the smallest, driest, lowest, flattest, and oldest. The 3-billion-year-old landscape has been worn down over time; Australia is the only continent without a tall mountain range. The Great Dividing Range (Australia's continental divide), close to the east coast, has the continent's highest mountains. The Australian Alps, on the southern end, provide winter recreation and summer runoff that irrigates the lowlands to the west. The further south one travels in Australia and New Zealand, the cooler it gets. The two main islands of New Zealand, which lie 1,200 mi. (1,920 km) to the southeast, are the opposite of flat, hot, and dry Australia. The green, hilly countryside is the product of a mild and moist climate. Some of its southern mountain peaks have a permanent snow cover.

The western two-thirds of Australia is a low plateau with the most deserts of any continent. Between this plateau and the fertile east coast's Great Dividing Range are dry lowlands called artesian basins. Artesian water is underground water, under pressure, that rises anywhere there is a crack in the earth's surface. Underground water percolates from the eastern mountains. Because of its high salt content, the water isn't drinkable and has only a limited crop application. But it can support the millions of sheep that dot the landscape. The westward-blowing trade winds (Plate 44) are stripped of their moisture by the Great Dividing Range, causing the rivers and lakes in central and western Australia to stay bone-dry most of the year (they are represented on the map by broken lines).

The Great Barrier Reef, the world's largest coral reef, runs 1,250 mi. (2,000 km) off the northeast coast of Australia. This multi-colored mecca for scuba divers is a collection of 2,500 individual reefs supporting thousands of species of marine life. The reefs are made from the skeletons of hundreds of species of sea coral. In order for reef building to take place, water temperature must exceed 65° F (18.3° C).

Australia and the other islands of Oceania are noted for their unusual wildlife. Because the lands have been separated from other continents for millions of years (Plate 2), many plants and animals are unique to this part of the world. Most of the world's marsupials (mammals that raise their immature young in external pouches), including the well-known kangaroo, live only in Australia. Two unusual nonmarsupials are the platypus and the echidna (the spiny anteater), the only mammals in the world that lay eggs. Only one placental mammal is a native everywhere in Oceania: the bat. No other mammal had the ability to cross the ocean. The dingo, a wild dog that has roamed Australia for thousands of years, originally came from Asia with the Aborigines. The emu and cassowary, large flightless birds which resemble ostriches, are also endemic to this continent. The koala, a leaf-eating, teddy bear-like marsupial, is a well-known resident of the tall eucalyptus trees. There are over 400 eucalyptus species and over 600 species of the acacia tree. These trees and other varieties of plant life unique to Australia are now flourishing in other parts of the world with similar climates, such as California.

## AUSTRALIA *

**Area:** 2,970,000 sq. mi. (7,692,300 km²). **Population:** 16,000,000. **Capital:** Canberra, 265,000. **Government:** Constitutional monarchy. **Language:** English. **Religion:** Protestant. **Exports:** Wool, iron ore, coal, bauxite, beef, cereals, and sugar. **Climate:** The tropical north gets heavy winter rains. The east coast is mild, and Tasmania is cool and damp. The rest of the country is very dry and seasonally hot. □ Australia, the smallest continent but one of the largest countries (about the size of the US without Alaska), has been called the "Land Down Under" because of its location south of the equator. The name " Australia" comes from the Latin "australis," meaning "southern." For many years the only European settlers were British and Irish immigrants. Australia was a huge, isolated English outpost in east Asian waters. After World War II, other Europeans were admitted. Immigration laws were later expanded to include Asian nationals; today Australia is a multiracial and multicultural nation. Almost a quarter of the population is foreign-born. Coastal parts of Australia were first explored by the Netherlands, Portugal, and Spain in the early 17th century. Captain Cook arrived and claimed the continent for Great Britain in 1770. The earliest immigrants were inmates in Britain's new overseas penal colony. The nation became completely independent in 1901. When the first Europeans arrived, approximately 300,000 Aborigines were living on the continent. Their ancestors migrated from Asia 20,000 to 50,000 years ago. These dark-skinned nomads were nearly wiped out by the Europeans' diseases and brutality. Their numbers are on the rise (currently 140,000) as the government is making restitution for the wrongs of the past. They are now able to retain their tribal lands and culture or to become thoroughly integrated into Australia's modern society. One well-known artifact, the boomerang, is an ancient Aborigine invention.

Even with growing industries, Australia's wealth continues to be derived from mining and agriculture. The country is the top exporter of bauxite, and it has large reserves of iron, coal, lead, zinc, and other minerals. Australia is the world's leading producer of wool. Most of the sheep are the incredibly wooly merinos, which thrive in the hot, dry interior ("outback"). Not surprisingly, the country is also the leading exporter of lamb and mutton. Cattle do well on the relatively barren ranches (called "stations"); the weather is mild and there is plenty of land for each animal. Australia is so large, and its meager population so scattered, that in the outback medical aid has to be rendered by airplane and children are taught at home by radio and mail correspondence.

Most Australians live in the cities. The populated areas, farms, and industries are located along the southeast coast in the state of *New South Wales.* Sydney (3,400,000), the largest city, has one of the world's great natural harbors. Sydney's ultra-modern opera house is a startling sight; its many pointed roofs resemble the sails of passing boats. Australia is made up of six states and two territories. The tiny *Australian Capital Territory,* located within New South Wales, is the site of the nation's capital, Canberra, which is the only major city that isn't on the coast. *Victoria,* an important farming state, is on the southeast coast. Its principal city and major port is Melbourne (2,850,000). The smallest state, *Tasmania,* is an island off the southeastern tip of the mainland. The Great Dividing Range submerges beneath the Bass Strait and surfaces to form the island. This important apple-growing region is a popular resort area. *Queensland* occupies the northeast corner of the nation. The savanna-like land is ideal for cattle raising. Three states—*Northern Territory, South Australia,* and *Western Australia*—make up two-thirds of the nation. Except for a narrow Mediterranean climate belt along the south coast, these states are mostly barren desert. Remarkable Ayers Rock, located near the center of Australia, may be the world's largest monolith. This enormous oval boulder is sacred to the Aborigines. It rises abruptly from the flat desert floor to a height of 1,140 ft. (348 m) and is nearly 2 mi. (3 km) long. It can be compared to an iceberg in that only about 5% of the rock is above ground.

## NEW ZEALAND

**Area:** 103,775 sq. mi. (268,777 km²). **Population:** 3,400,000. **Capital:** Wellington, 350,000. **Government:** Constitutional monarchy. **Language:** English; Maori. **Religion:** Protestant. **Exports:** Dairy products, lamb, wool, fruit, fish, and paper products. **Climate:** Mild; temperatures vary with latitude and altitude. □ Mild weather and ample rainfall are two of the factors contributing to New Zealand's fine pasturelands; the third is fertilization. New Zealand is the world's largest exporter of dairy products and lamb and the second largest producer of wool. Because its orchards are harvested when the northern hemisphere is in winter, New Zealand produce is a valuable export. Almost all shipping passes through the nation's largest city, Auckland (825,000), on the North Island which is home to three-fourths of the population.

New Zealand was discovered by Dutch explorer Abel Tasman in 1642. He was driven off by the Maoris, who had arrived from Polynesia around 700 years earlier. It wasn't until Captain Cook established good relations with the natives in 1769 that Britain was able to settle the islands. By the early 19th century, the Maoris thought that immigration had gotten out of hand and began attacking the Europeans. British troops arrived to establish order, and eventually a peace treaty was signed in 1840, giving the British sovereignty over the islands but assuring land ownership to the Maoris. Though the exact interpretation of the treaty is still being debated, the Maoris are once again thriving because of the great effort being made to respect their civil and property rights. The Maoris make up about 10% of the population, and intermarriage between these Polynesians and New Zealanders of European descent is common. New Zealand has always been a politically progressive nation and was one of the first to enact social welfare legislation. In 1893, New Zealand's women were the first ones to receive full voting rights.

Because New Zealand is located at the point where the Pacific tectonic plate passes under the Indo-Australian plate (Plate 2), the islands are geologically very active. In the center of North Island is a barren plateau with a hellish character: active volcanoes, steaming fumaroles (vapor vents), powerful geysers, boiling hot springs, and bubbling mud pools. The world's first steam-powered, geothermal plant was built here in 1961. Rushing rivers also provide hydroelectric power. The tallest mountains are on rugged South Island. Its southwest coast is called Fiordland. The most remarkable fiord (the English spell it with an "i") is Milford Sound, whose mile-high sea cliffs are the world's tallest. In these waters, an unusual combination of low light, salt water, fresh water, and warm water temperatures create a unique sea ecology that contains the world's largest formation of black coral. New Zealand does not have any snakes, but a native reptile called a tuatara predates the dinosaurs. Moas were enormous birds (up to 12 ft., 3.7 m tall); they were hunted into extinction by the early Maori tribes. The kiwi is not just a popular fruit; it is also a wingless bird. New Zealanders have this bird (the national symbol) in mind when they refer to themselves as Kiwis.

## PAPUA NEW GUINEA

**Area:** 178,800 sq. mi. (463,092 km²). **Population:** 4,150,000. **Capital:** Port Moresby, 130,000. **Government:** Constitutional monarchy. **Language:** English; Pidgin; nearly 700 dialects. **Exports:** Copper, coffee, cocoa, copra, and timber. **Climate:** Tropical and damp. □ Papua New Guinea includes the eastern half of the island of New Guinea, the Bismarck Archipelago, Bougainville and Buka in the Solomon Islands, and about 600 smaller islands. Some of the world's most primitive peoples live in this country, parts of which remain unexplored. The dark-skinned Melanesian inhabitants represent a great number of cultures and speak over 700 dialects. Pidgin English, which is English heavily flavored by native dialects, is the closest thing to a common tongue. In the late 19th century, the northern part of the island was controlled by Germany, which gave it to Australia following World War I. In 1975, Papua New Guinea received its independence, but continues to rely on Australian aid.

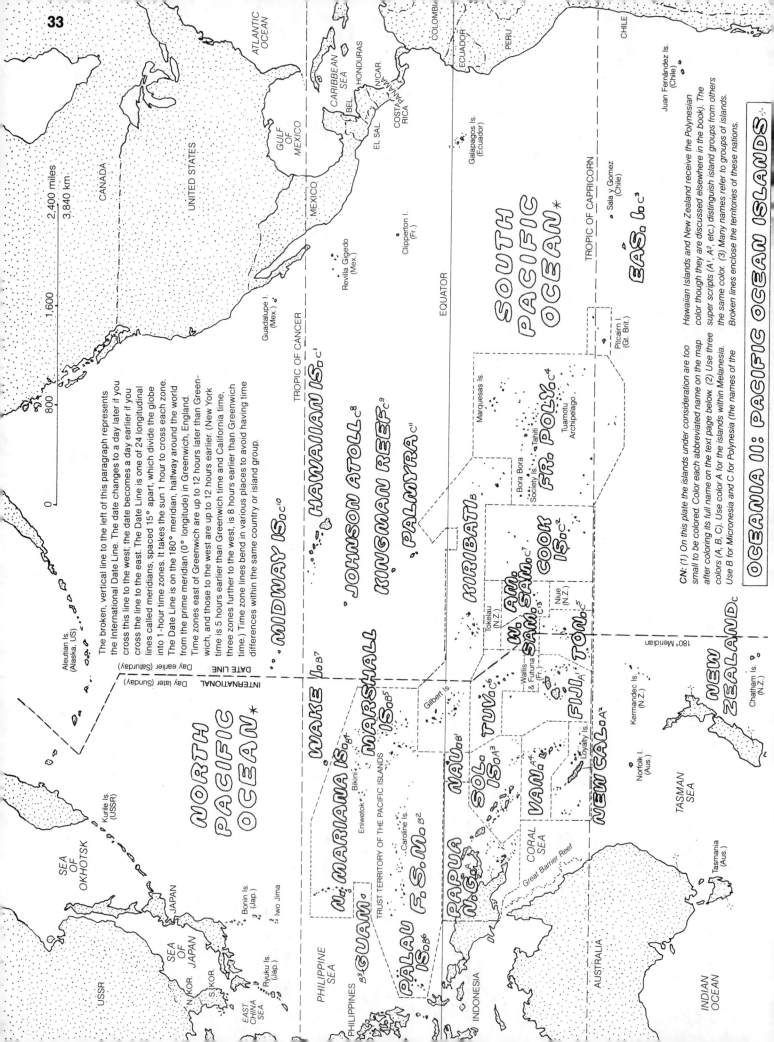

# 33

# OCEANIA II: PACIFIC OCEAN ISLANDS

The broken, vertical line to the left of this paragraph represents the International Date Line. The date changes to a day later if you cross this line to the west; the date becomes a day earlier if you cross the line to the east. The Date Line is one of 24 longitudinal lines called meridians, spaced 15° apart, which divide the globe into 1-hour time zones. It takes the sun 1 hour to cross each zone. The Date Line is on the 180° meridian, halfway around the world from the prime meridian (0° longitude) in Greenwich, England. Time zones east of Greenwich are up to 12 hours later than Greenwich, and those to the west are up to 12 hours earlier. (New York time is 5 hours earlier than Greenwich time and California time, three zones further to the west, is 8 hours earlier than Greenwich time.) Time zone lines bend in various places to avoid having time differences within the same country or island group.

CN: (1) On this plate the islands under consideration are too small to be colored. Color each abbreviated name on the map after coloring its full name on the text page below. (2) Use three colors (A, B, C). Use color A for the islands within Melanesia. Use B for Micronesia and C for Polynesia (the names of the Hawaiian Islands and New Zealand receive the Polynesian color though they are discussed elsewhere in the book). The super scripts (A[1], A[2], etc.) distinguish island groups from others the same color. (3) Many names refer to groups of islands. Broken lines enclose the territories of these nations.

INTERNATIONAL DATE LINE

Day later (Sunday) / Day earlier (Saturday)

NORTH PACIFIC OCEAN *

SOUTH PACIFIC OCEAN *

ATLANTIC OCEAN

INDIAN OCEAN

EQUATOR

TROPIC OF CANCER

TROPIC OF CAPRICORN

180° Meridian

2,400 miles
3,840 km

0   800   1,600

CANADA

UNITED STATES

MEXICO

GULF OF MEXICO

CARIBBEAN SEA

BEL. / HONDURAS / NICAR. / COSTA RICA / EL SAL.

COLOMBIA

ECUADOR

PERU

CHILE

Galapagos Is. (Ecuador)

Clipperton I. (Fr.)

Revilla Gigedo (Mex.)

Guadalupe I. (Mex.)

Juan Fernández Is. (Chile)

Sala y Gomez (Chile)

EAS₍ₐ₎ I₍ₐ₎ C³

Pitcairn I. (Gt. Brit.)

MIDWAY IS₍ₐ₎C¹⁰

HAWAIIAN IS₍ₐ₎C¹

JOHNSON ATOLL C⁸

KINGMAN REEF C⁹

PALMYRA C¹¹

WAKE I₍ₐ₎ B⁷

N. MARIANA IS₍ₐ₎B⁵

MARSHALL IS₍ₐ₎B⁵

Bikini / Eniwetok

Caroline Is.

GUAM B³

PALAU IS₍ₐ₎B⁶

F.S.M.₍ₐ₎B²

TRUST TERRITORY OF THE PACIFIC ISLANDS

KIRIBATI B

Gilbert Is.

NAU₍ₐ₎B¹

Marquesas Is.

Society Is. / Bora Bora / Tahiti

FR. POLY.₍ₐ₎C⁴

Tuamotu Archipelago

COOK IS₍ₐ₎C²

Niue (N.Z.)

Tokelau (N.Z.)

W. AM. SAM.₍ₐ₎C⁷ / SAM.₍ₐ₎C¹³

Wallis & Futuna (Fr.)

TON'GA₍ₐ₎C⁵

TUV₍ₐ₎C⁶

FIJI₍ₐ₎ C¹

SOL. IS₍ₐ₎A³

VAN₍ₐ₎A⁴

NEW CAL.₍ₐ₎A²

PAPUA N. G.₍ₐ₎A¹

Loyalty Is.

Norfolk I. (Aus.)

Kermandec Is. (N.Z.)

Chatham Is. (N.Z.)

NEW ZEALAND C

TASMAN SEA

AUSTRALIA

Tasmania (Aus.)

Great Barrier Reef

CORAL SEA

INDONESIA

PHILIPPINES

PHILIPPINE SEA

JAPAN

N. KOR. / S. KOR.

EAST CHINA SEA

SEA OF JAPAN

SEA OF OKHOTSK

USSR

Ryukyu Is. (Jap.)

Bonin Is. (Jap.) / Iwo Jima

Kurile Is. (USSR)

Aleutian Is. (Alaska, US)

The islands of Oceania, most of which are located in the southwest Pacific, are grouped according to regions: Melanesia, Micronesia, and Polynesia. Melanesia ("black islands") refers to the skin color of the inhabitants of New Guinea and some of the islands to the southeast. Micronesia ("small islands") describes the size of the islands that lie to the north of Melanesia.

The islands of Oceania are coral reefs or the peaks of submerged volcanoes. The latter are prone to earthquakes and eruptions. Their moisture-capturing peaks make these islands wetter than the low-lying reefs. The soil of the older volcanic islands, such as Hawaii, is enriched by the gradual breakdown of volcanic rock. Coral reefs, formed by the skeletons of sea coral, have a thin soil which is marginally fertile. Most reefs are in the form of an atoll: a ring of land surrounding a lagoon. The atoll is created when coral forms a fringe around the base of a volcanic peak which eventually sinks beneath the sea, leaving in its place a lagoon surrounded by the coral.

The Pacific, the world's largest and deepest ocean, measures half the distance around the globe. The ocean isn't always as benign as its name implies. The typhoons (cyclones) of the western Pacific inflict terrible destruction. These storms are generally larger and more powerful than the hurricanes of the eastern Pacific and the Atlantic Ocean. Except for seasonal storms and the heat and humidity of the equatorial waters, Oceania usually enjoys a balmy climate cooled by the trade winds.

The islands of Oceania have been part of a romantic Western myth that pictures the "South Seas" as an island paradise with gleaming white beaches, swaying palms, crystal-clear waters, and attractive natives pursuing a carefree existence filled with food, song, and dance. This vision may have had some truth prior to the arrival of the Europeans in the 16th century, but in the years that followed most native societies were forever changed by colonization, disease, Christian missionaries, commercial exploitation, and military battles. European and Asian immigrants, who on some islands are in the majority, have transformed the traditional village culture that easily lived within its environment into dependent modern societies. Many islands have acquired their independence in the second half of the 20th century. Others remain in an indefinite trust status (the US has many such islands).

## MELANESIA A-1

The residents of the "black islands," along with the Australian Aborigines, are the darkest-skinned inhabitants of Oceania. These short, powerfully built, curly-haired people probably descend from the earliest migrants from Asia. Some of the tribes still live in the stone age. Status within a Melanesian tribe is acquired by the accumulation or exchange of material goods.

### FIJI A-1
Area: 7,075 sq. mi. (18,324 km²). Population: 720,000. Capital: Suva, 69,000. Government: Republic. Language: English; Fijian; Hindi. Religion: Christianity 50%, Hinduism 40%. Exports: Sugar, copra, gold, and oil. □ Two large volcanic islands make up 87% of the land area of Fiji (fee'jee). The 800 remaining islands are mostly low-lying coral atolls. Prosperous descendants of 19th century Indian immigrants from India outnumber the landowning Melanesians.

### NEW CALEDONIA A-2
Area: 7,380 sq. mi. (19,114 km²). Population: 155,000. Capital: Noumea, 56,100. Government: French Overseas Territory. Language: French, native dialects. Religion: Roman Catholic 60%, Protestant 30%. Exports: Nickel, chrome, iron, cattle, and coffee. □ New Caledonia, which includes the Loyalty Islands, was once the world's largest producer of nickel. French settlers oppose the Melanesian minority's demand for independence.

### SOLOMON ISLANDS A-3
Area: 11,504 sq. mi. (29,795 km²). Population: 335,000. Capital: Honiara, 16,500. Government: Constitutional monarchy. Language: Pidgin; English; native dialects. Religion: Protestant 55%, Roman Catholic 20%. Exports: Fish, bananas, timber, and copra. □ In the Solomons live some of the most

primitive tribes in Oceania. In World War II, the US won a major naval victory over the Japanese in the Coral Sea. Fierce fighting occurred on Guadalcanal.

### VANUATU A-4
Area: 5,700 sq. mi. (14,763 km²). Population: 145,000. Capital: Port Vila, 10,200. Government: Republic. Language: Bislama, English, French. Religion: Christianity. Exports: Copra. □ The 80 mountainous islands of Vanuatu (va noo' a too) were called the New Hebrides when England and France jointly governed them. Bislama, the most commonly used of over 100 spoken languages, is a form of Pidgin.

## MICRONESIA B-1

The "small islands" of Micronesia lie to the north of Melanesia. Though generally taller and lighter-skinned than Melanesians, Micronesians do not have any distinguishing physical characteristics. They tend to resemble their closest neighbors—Asians, Melanesians, or Polynesians. Leadership in the remaining traditional societies of Micronesia (and Polynesia) is based on royal lineage and inherited rank.

### KIRIBATI B
Area: 300 sq. mi. (777 km²). Population: 70,000. Capital: Tarawa, 1,825. Government: Republic. Language: Kiribati; English. Religion: Roman Catholic 45%, Protestant 40%. Exports: Copra, produce, and phosphates. □ Kiribati (kir'ih bans) is a nation of 33 small islands (including the Gilberts) in an area nearly the width of the US. The economy relies on aid from Britain.

### NAURU B-1
Area: 8 sq. mi. (21 km²). Population: 9,000. Capital: Yaren district. Government: Republic. Language: Nauruan; English. Religion: Protestant 65%, Roman Catholic 30%. Exports: Phosphates. □ Nauru (nah' roo) is the third smallest country in the world (after Vatican City and Monaco) and one of the richest, per capita. The island is a plateau of practically pure phosphate. In anticipation of the eventual depletion of reserves, revenues are being wisely invested. Nauru is suing Australia for exploiting this resource when it was the administrator of the island.

### US TERRITORIES B-2–B-7
In 1947, the United Nations placed more than 2,000 islands, formerly occupied by the Japanese, under the Trust Territory of the Pacific Islands to be administered by the United States. These islands were: the Federated States of Micronesia (FSM), formerly the Caroline Islands; the Marshall Islands; the Northern Mariana Islands; and the Palau Islands. Only the latter is still a Trust Territory; the others have been granted self-governing status in "free association" with the US. Under this arrangement, the US continues to give economic aid to all former Trust Territories and handles their defense. The Northern Marianas have US commonwealth status, comparable to that of Puerto Rico. Another commonwealth in Micronesia is Guam, the most populous (120,000) of the US islands territories, which was acquired (along with the Philippines and Puerto Rico) from Spain in 1899. Guam and the Northern Mariana chain are vacation spots for the Japanese. Wake Island is administered by the US Air Force. Bikini and Eniwetok atolls in the Marshall Islands were US nuclear test sites; after 40 years, Bikini is still uninhabitable.

## POLYNESIA C-1
Most of the "many islands" of Polynesia lie east of the International Date Line in the central Pacific. Polynesians, believed to be of Malay ancestry, are the tallest and have the lightest skin color of the natives of Oceania. Residents of these far-flung Pacific islands were the last migrants in the movement from Southeast Asia. The Polynesians were superb sailors, capable of accurately navigating long distances without the aid of instruments. New Zealand (Plate 32) and Hawaii (Plate 9), are the largest islands in Polynesia.

### COOK ISLANDS C-2
Area: 92 sq. mi. (238 km²). Population: 20,500. Capital: Avarua, 9,550. Gov-

ernment: Self-governing citizens of New Zealand. Language: English; Maori dialect. Religion: Protestant. Exports: Copra and fruit. □ These 15 islands are named after the British sea captain James Cook, who explored and charted much of Oceania. The natives are related to the Maoris of New Zealand.

### EASTER ISLAND C-3
Area: 46 sq. mi. (119 km²). Population: 2,100. Capital: Hanga Roa. Government: Territory of Chile. Language: Spanish. Religion: Roman Catholic. Exports: Wool. □ The island was named on Easter Sunday. No one has unraveled the mystery of more than 600 huge heads, ranging in size from 12 to 40 ft. (3.7–12.2 m), carved out of volcanic rock by an ancient civilization.

### FRENCH POLYNESIA C-4
Area: 1,500 sq. mi. (3,885 km²). Population 145,000. Capital: Papeete, 24,000. Government: French Overseas Territory. Language: French; Tahitian. Religion: Protestant 50%, Roman Catholic 35%. Exports: Copra, sugar, fruit, and mother-of-pearl. □ The Society Islands and the Marquesas are among the 130 islands of French Polynesia. The handsome natives of Tahiti, the largest island, were immortalized by the French painter Paul Gauguin.

### TONGA C-5
Area: 300 sq. mi. (777 km²). Population: 110,000. Capital: Nukualofa, 18,350. Government: Constitutional monarchy. Language: Tongan; English. Religion: Christianity. Exports: Copra and fruit. □ The "Friendly Islands" were so named by Captain Cook for his warm welcome. Tonga's 150 islands make up Polynesia's last kingdom. The people are very large and heavyset—the present King weighs well over 400 lbs. A weak economy depends upon paychecks sent home by Tongans working in New Zealand.

### TUVALU C-6
Area: 10 sq. mi. (25 km²). Population: 7,300. Capital: Funafuti, 2,200. Government: Constitutional monarchy. Language: Tuvaluan, English. Religion: Christianity. Exports: Copra. □ Tuvalu (formerly the Ellice Islands) is the world's fourth smallest nation in size and second smallest in population. There were three times as many people living on Tuvalu in the 19th century, but the population was decimated by "blackbirders," Europeans who seized natives to work plantations on other islands.

### US TERRITORIES C-7–C-11
American Samoa, Jarvis Island, Johnson Atoll, Kingman Reef, Midway Islands, and Palmyra. About 85,000 people (more than twice the present population) have moved from American Samoa to the US mainland. Both of the islands of Midway are administered by the US Navy, the turning point of World War II in the Pacific. These territories have taken on new importance with the recent enlargement of US coastal waters. In 1983, the US followed other coastal nations and proclaimed a 200 mi. (320 km) "exclusive economic zone." In 1988, the US widened the zone of "total sovereignty" from 3 mi. (4.8 km) to 12 mi. (19 km). Foreign ships may not enter the 200-mile zone except in free navigation or to lay seabed cables, and they cannot operate within 12 mi. of a US shoreline for any reason. This policy means that any US-owned island, no matter how small, can control all fishing and mineral rights within 200 miles. A "wet America" has been created equal in size to continental America. In the absence of international agreements, such exclusive economic zones present opportunities for the unrestricted exploitation of the oceans.

### WESTERN SAMOA C-12
Area: 1,098 sq. mi. (2,844 km²). Population: 190,000. Capital: Apia, 33,800. Government: Constitutional monarchy. Language: Samoan; English. Religion: Christianity. Exports: Copra and bananas. □ The people of these four islands are the tallest and most powerfully built in Oceania. Western Samoa, the first Polynesian state to become independent (1962), was a colony of Germany before World War I, but it would be difficult to find any Teutonic influence on the easygoing, music and dance-filled life-style of these attractive people.

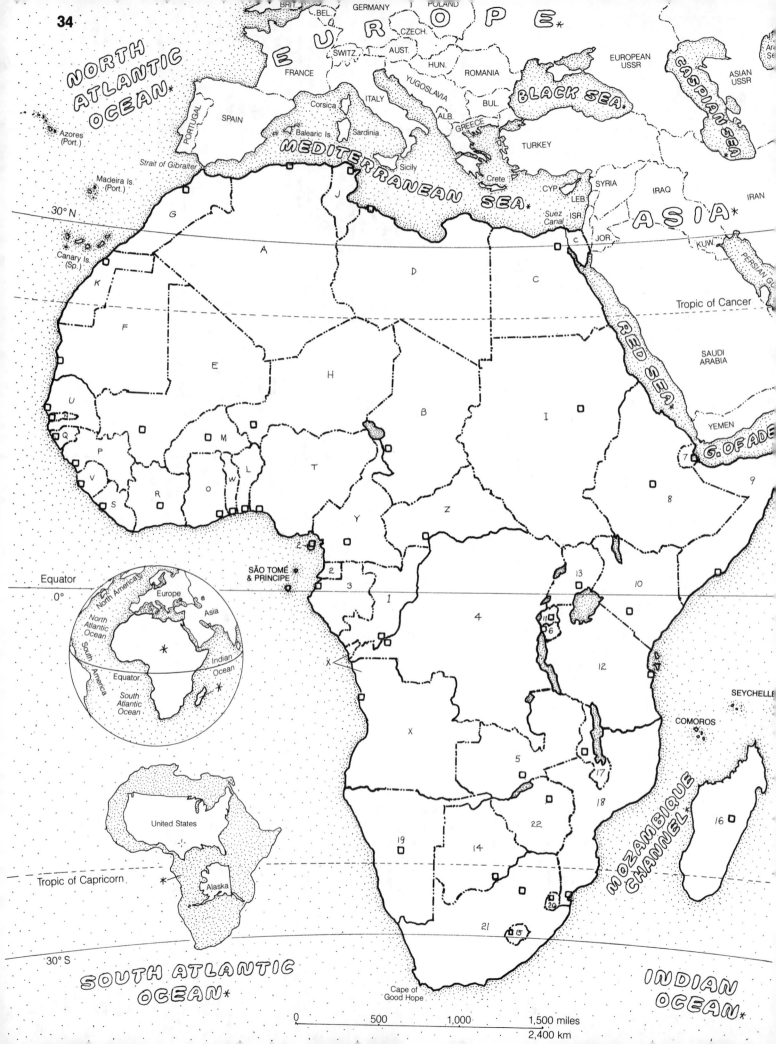

# 34

# AFRICA: THE COUNTRIES

CN: (1) Color a country, then its name. (2) Use light colors on the map of colonial Africa (23-29) so that you can see the national boundaries. (3) The island nations São Tomé and Principe, Comoros, and the Seychelles are too small to color. The islands of Cape Verde (Plate 37) are not shown.

Africa covers 11,700,000 sq. mi. (30,279,600 km²); it is the second largest continent (after Asia). It has 53 nations and 600 million people—the third largest population (after Asia and Europe). The continent is divided into two racial and cultural zones by the Sahara Desert. Northern nations, bordering the Mediterranean, are populated by light-skinned, Arabic-speaking Muslims. Countries south of the Sahara are mostly populated by black Africans who speak hundreds of different languages (many are Bantu dialects). The most common form of Bantu is Swahili, which is the lingua franca of eastern Africa. Muslim missionaries and their Christian counterparts are making significant inroads among the native religions (mostly animist) of black Africans. Islam is gaining because blacks regard it as an African religion despite its Middle Eastern origins; Christianity, with similar origins, is considered European.

Though the Portuguese began establishing coastal colonies as early as the 15th century, it wasn't until the late 1800s that Europeans penetrated the African interior and began carving up the continent in earnest. At the outbreak of World War I (1914), only Liberia and most of Ethiopia remained free of foreign domination. The second half of the 20th century brought a great rush toward independence, and colonialism formally came to an end in 1990 with the free elections in Namibia. France, which had the largest African empire, maintains close relations with most of its former colonies, and many are dependent upon it for economic and military aid.

Independence has not meant freedom for most Africans; they are generally ruled by military or one-party governments. The boundaries of the new countries are virtually the same as those drawn by the colonialists, who were either ignorant of, or indifferent to, traditional tribal divisions. Many African nations suffer from such arbitrary borders that separate related groups or confine traditional enemies within the same country. Tribal loyalties often take precedence over loyalty to the new nation.

The problems facing these young nations are enormous. Fertile land and rainfall are not generally plentiful in Africa. In the few places where agriculture is productive, gains have been erased by excessive population growth. The largest farms still follow the colonial practice of growing cash crops for export instead of food for local consumption. When world commodity prices are depressed, cash crops do not provide enough income to buy food. Most African nations are dependent on declining foreign aid. Except for a few oil producers, the nations rich in natural resources have been unable to profitably mine and market them. In many cases, potentially healthy economies have been wrecked by communist mismanagement. Others have been victimized by brutal and corrupt leaders who squandered precious revenues on ill-conceived public works projects or monuments of self-aggrandizement. Still other countries have been torn apart by civil war. Hunger, poverty, disease, and illiteracy are on the increase in many nations across the continent.

In the late 1980s, Western interest (and economic aid) shifted from Africa to the emerging democracies of eastern Europe. With the "cold war" winding down, another major source of aid was drying up—African nations could no longer receive assistance by playing one side against the other. The one commodity the industrialized world (particularly Europe) seems anxious to send to Africa is toxic waste. Though most nations have stopped accepting shipments, others are so desperate they will not reject revenue from any source.

## NORTHERN
ALGERIA A ALGIERS
CHAD B N'DJAMENA
EGYPT C CAIRO
LIBYA D TRIPOLI
MALI E BAMAKO
MAURITANIA F NOUAKCHOTT
MOROCCO G RABAT
NIGER H NIAMEY
SUDAN I KHARTOUM
TUNISIA J TUNIS
WESTERN SAHARA K AAIÚN

## WESTERN
BENIN L PORTO NOVO
BURKINA FASO M OUAGADOUGOU
GAMBIA N BANJUL
GHANA O ACCRA
GUINEA P CONAKRY
GUINEA-BISSAU Q BISSAU
IVORY COAST R YAMOUSSOUKRO
LIBERIA S MONROVIA
NIGERIA T LAGOS
SENEGAL U DAKAR
SIERRA LEONE V FREETOWN
TOGO W LOMÉ

## CENTRAL
ANGOLA X LUANDA
CAMEROON Y YAOUNDÉ
CENTRAL AFRICAN REP. Z BANGUI
CONGO 1 BRAZZAVILLE
EQUATORIAL GUINEA 2 MALABO
GABON 3 LIBREVILLE
ZAIRE 4 KINSHASA
ZAMBIA 5 LUSAKA

## EASTERN
BURUNDI 6 BUJUMBURA
DJIBOUTI 7 DJIBOUTI
ETHIOPIA 8 ADDIS ABABA
SOMALIA 9 MOGADISHU
KENYA 10 NAIROBI
RWANDA 11 KIGALI
TANZANIA 12 DAR ES SALAAM
UGANDA 13 KAMPALA

## SOUTHERN
BOTSWANA 14 GABORONE
LESOTHO 15 MASERU
MADAGASCAR 16 ANTANANARIVO
MALAWI 17 BAMAKO
MOZAMBIQUE 18 MAPUTO
NAMIBIA 19 WINDHOEK
SWAZILAND 20 MBABANE
SOUTH AFRICA 21 PRETORIA
ZIMBABWE 22 HARARE

## COLONIAL AFRICA

Except for Liberia and Ethiopia, the entire continent was under European control by the early 19th century.

BELGIAN 23
BRITISH 24
FRENCH 25
GERMAN 26
ITALIAN 27
PORTUGUESE 28
SPANISH 29

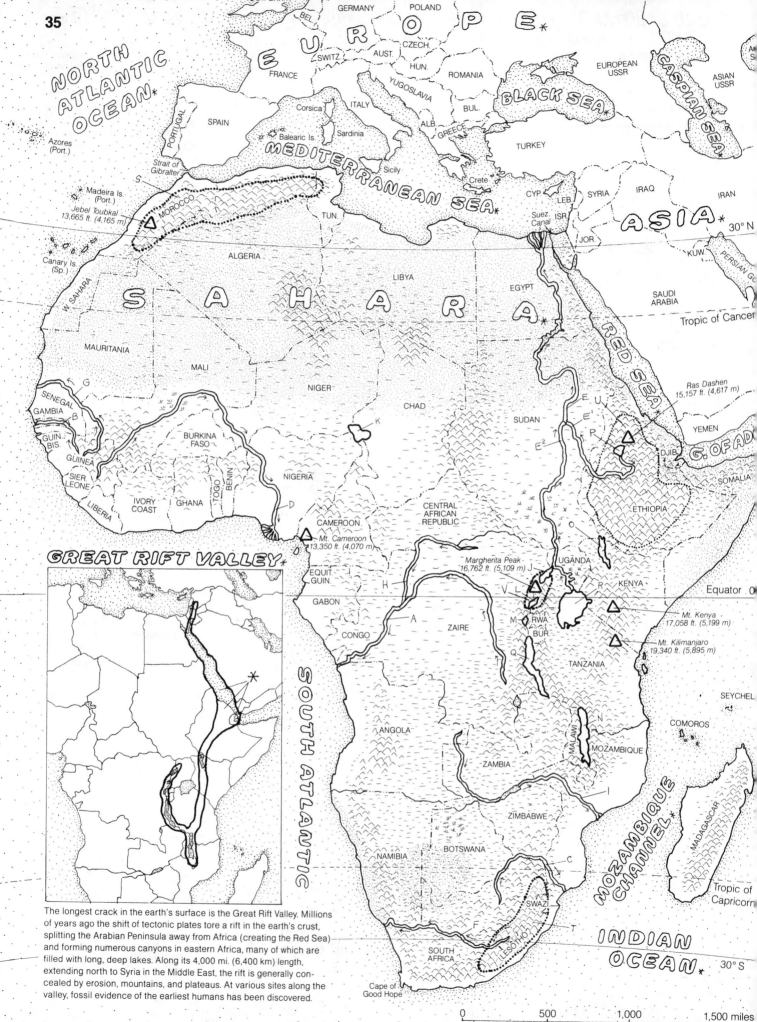

# 35

NORTH ATLANTIC OCEAN ✳

EUROPE ✳

GT. BRIT.
GERMANY
POLAND
BEL.
CZECH.
FRANCE
SWITZ.
AUST.
HUN.
ROMANIA
YUGOSLAVIA
BUL.
ITALY
ALB.
GREECE
Corsica
Balearic Is.
Sardinia
Sicily
Crete

BLACK SEA ✳

EUROPEAN USSR

CASPIAN SEA ✳

ASIAN USSR

TURKEY

MEDITERRANEAN SEA ✳

PORTUGAL
SPAIN

Azores (Port.)

Madeira Is. (Port.)
Jebel Toubkal 13,665 ft. (4,165 m)
S
MOROCCO

Strait of Gibralter

Canary Is. (Sp.)

30° N

CYP.
LEB.
SYRIA
IRAQ
IRAN
Suez Canal
ISR.
JOR.
ASIA
KUW.
PERSIAN G.

TUN.

W. SAHARA
ALGERIA
LIBYA
EGYPT

SAHARA

SAUDI ARABIA

Tropic of Cancer

RED SEA ✳

Ras Dashen 15,157 ft. (4,617 m)

MAURITANIA
MALI
NIGER
CHAD
SUDAN

YEMEN

G. OF AD.

SENEGAL
G
GAMBIA
B
GUIN. BIS.
GUINEA
SIER. LEONE
LIBERIA
IVORY COAST
BURKINA FASO
GHANA
TOGO
BENIN
NIGERIA
K

E
U
E¹
P
E²
DJIB.
SOMALIA
ETHIOPIA

D
CAMEROON
Mt. Cameroon 13,350 ft. (4,070 m)
D

CENTRAL AFRICAN REPUBLIC

O

EQUIT. GUIN.
GABON
CONGO
H
A
ZAIRE

UGANDA

Margherita Peak 16,762 ft. (5,109 m) J
V
L
M
RWA.
BUR.
Q

R
KENYA

Equator 0

Mt. Kenya 17,058 ft. (5,199 m)
Mt. Kilimanjaro 19,340 ft. (5,895 m)

TANZANIA

SEYCHEL.

COMOROS

ANGOLA
ZAMBIA
MALAWI
N
MOZAMBIQUE

SOUTH ATLANTIC

MOZAMBIQUE CHANNEL ✳

MADAGASCAR

ZIMBABWE
I

NAMIBIA
BOTSWANA
C

SWAZI
F
LESOTHO
T
SOUTH AFRICA

Tropic of Capricorn

INDIAN OCEAN ✳

30° S

Cape of Good Hope

GREAT RIFT VALLEY ✳

The longest crack in the earth's surface is the Great Rift Valley. Millions of years ago the shift of tectonic plates tore a rift in the earth's crust, splitting the Arabian Peninsula away from Africa (creating the Red Sea) and forming numerous canyons in eastern Africa, many of which are filled with long, deep lakes. Along its 4,000 mi. (6,400 km) length, extending north to Syria in the Middle East, the rift is generally concealed by erosion, mountains, and plateaus. At various sites along the valley, fossil evidence of the earliest humans has been discovered.

0     500     1,000     1,500 miles
2,400 km

## PRINCIPAL RIVERS:
CONGO (ZAIRE)ₐ → CONGO (ZAIRE)A
GAMBIA B
LIMPOPO C
NIGER D
NILE E
　　BLUE NILE E'
　　WHITE NILE E²
ORANGE F
SENEGAL G
UBANGI H
ZAMBEZI I

## PRINCIPAL LAKES:
L. ALBERT J
L. CHAD K
L. EDWARD L
L. KIVU M
L. NYASA (MALAWI) N
L. RUDOLPH (TURKANA) O
L. TANA P
L. TANGANYIKA Q
L. VICTORIA R

## PRINCIPAL MOUNTAIN RANGES:
ATLAS MTS. S
DRAKENSBERG T
ETHIOPIAN HIGHLANDS U
RUWENZORI V

*CN: (1) On the large map, use gray on the triangles representing important mountain peaks. (2) Use gray for the map of the Great Rift Valley on the far left. (3) Use light colors on the map of the land regions below.*

Over 80% of Africa lies between the tropics of Cancer and Capricorn; it thus has the largest tropical region of any continent. It has been said that the night hours are Africa's winter. But not all of Africa is warm; there are glacial areas in some eastern mountain ranges located at the equator. Curiously, in Africa the equatorial regions are not the hottest—it actually gets hotter further from the equator (except for the Mediterranean and southernmost coasts, which have pleasant climates). Though most of the continent is very dry to semiarid, heavy rainfall occurs in the equatorial regions, particularly in central and western Africa.

The African landscape has relatively little fertile territory: topsoil is generally thin, the deserts are huge, and most of the wetter regions are covered by a thick jungle. Tree-dotted semiarid grasslands (savannas) occupy wide areas of the continent and support an enormous population of large animals: elephants, giraffes, rhinoceroses, lions, and others. In the rain forests of central Africa, many animals live in tall trees, high above the dark, dank jungle floor: monkeys, chimpanzees, gorillas, reptiles, and birds.

Africa is a plateau made of ancient rock. It is rimmed by narrow coastal lowlands. Most of the mountain ranges are in the eastern and southern portions, in high Africa. Here the plateau reaches an altitude of 6,000 ft. (1,830 m) and then slopes even higher to form the Drakensberg Mountains, which tower over the coast of southern Africa. The most fascinating mountains are in the glacier-covered Ruwenzori Range between Lakes Edward and Albert on the Zaire-Uganda border. There is an almost constant cloud cover, so the sight of the glacier-covered peaks, nine of which reach over 16,000 ft. (4,878 m), is a rare and impressive experience. Ordinary plants have been known to grow to extraordinary sizes on the Ruwenzori slopes because of unusually favorable conditions. To the east lies snow-capped Kilimanjaro, Africa's tallest mountain (19,340 ft., 5,895 m); it is one of a group of rift-formed volcanic peaks.

One reason Africa was the last major continent to be explored and colonized by Europeans was that it presented formidable physical obstacles: an unusually smooth coastline with few peninsulas, islands, and natural harbors; a forbidding interior of deserts, jungles, and hot, arid plains; and a shortage of navigable rivers. Most African rivers, including the four major ones (the Nile, the Congo or Zaire, the Niger, and the Zambezi), are interrupted by impassable rapids and waterfalls. The Nile (including the White Nile) is the world's longest river at 4,150 mi. (6,640 km). Lake Victoria is credited as its source, but most of the White Nile (whose waters are pale green) is dissipated in the swamps of southern Sudan. Nearly 90% of the water that flows along the main Nile through Egypt comes from Lake Tana, Ethiopia, via the shorter Blue Nile (whose waters are blue). The massive Congo, locally called the Zaire, is the world's second largest river by volume (after the Amazon) and is 2,600 mi. (4,160 km) long. The Niger River is unusual in that it travels nearly as far to reach the sea, even though it begins only 150 mi. (240 km) from the coast. The major lakes are found in the Great Rift Valley (Lake Victoria, the world's second largest after Lake Superior, is actually situated on a plateau between two arms of the valley). Extremely deep Lake Tanganyika, on the Zaire-Tanzania border, is the world's longest (420 mi., 680 km).

The dominant geographical feature of Africa is the world's largest desert, the constantly expanding Sahara, which is currently the size of the continental United States. Rainfall there is scant and unpredictable. The only available water in the "land of thirst" is found in isolated oases and in the Nile River on its eastern edge. Yet as recently as 5,000 years ago the Sahara (Arabic for "emptiness") was a grassland. Today it is covered mostly by rock, gravel, and salt deposits. Sand dunes account for only one-fifth of the desert's surface. Because roads are so few, the camel ("ship of the desert") remains the most reliable form of transportation. The camel's heavy-lidded eyes and closeable nostrils enable it to withstand the fierce sandstorms that can turn day into night as they cut swaths as wide as 300 mi. (480 km). Dust from the Sahara can blow as far north as the Swiss Alps. In the summer, desert winds bring intense heat to the Mediterranean region. The world's highest shade temperature, 136° F (58° C), was recorded near the Libyan coast. The Namib Desert, along the coast of Namibia in southwest Africa, has the world's tallest sand dunes: some reach 1,000 ft. (305 m). Fog from the adjacent ocean provides its only moisture. The cold ocean currents that prevent rain from reaching the shore are similar to the conditions creating the deserts along the west coast of South America (Plate 17).

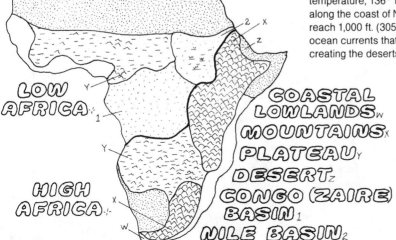

LOW AFRICA :
HIGH AFRICA :

COASTAL LOWLANDS W
MOUNTAINS X
PLATEAU Y
DESERT Z
CONGO (ZAIRE) BASIN 1
NILE BASIN 2

## LAND REGIONS:
The dark line across this smaller map divides low Africa from the higher lands of the east and south, known as high Africa. The Congo (Zaire) Basin and the lowlands of west Africa are covered with rain forests. Surrounding these jungle areas are broad semi-arid plateaus, mostly covered by savannas (grasslands). Still further to the north, south, and east lie Africa's deserts.

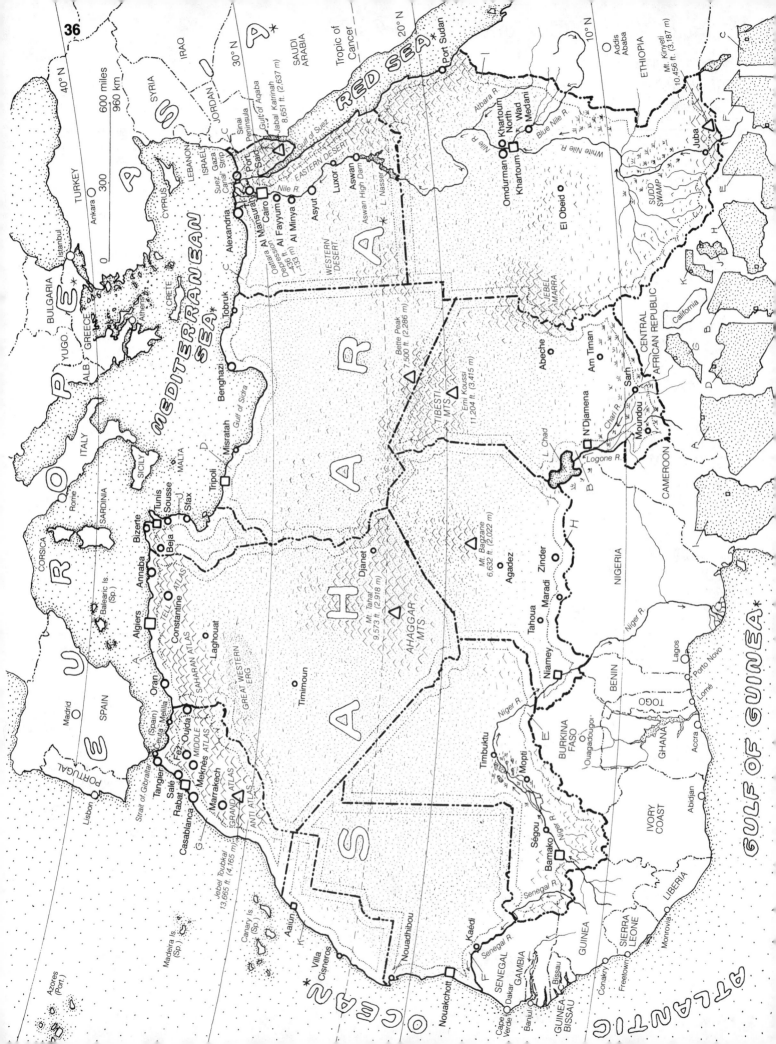

# AFRICA: NORTHERN

Northern Africa is set apart from the rest of the continent by the Sahara Desert and by the Islamic culture. The people of this region are light-skinned Arabs, Berbers, Moors (a mixture of the two), and Egyptians. Their homelands are almost entirely desert, except for the northern coast, where most of the population lives. Mediterranean crops, including grapes, olives, citrus, and dates, are grown in abundance. The Sahara is expanding to the south and west, covering the northern portions of sub-Sarahan Mauritania, Mali, Niger, Chad, and the Sudan. The nomadic residents of the desert regions are generally light-skinned Muslims who have little in common with the black populations in the southern parts of those nations. These sub-Saharan countries, all under military rule, are the poorest in Africa. The three in the middle—Mali, Niger, and Chad—are landlocked. They have few natural resources and must depend upon foreign aid and the limited agriculture of small but fertile regions along their southern borders. Frequent droughts have added to their misery. In the early 20th century, French West Africa included all of northern Africa except Libya (which was controlled by Italy), Western Sahara and part of Morocco (Spain); Egypt (Britain); and Sudan (Britain and Egypt).

## ALGERIA A

**Area:** 952,580 sq. mi. (2,467,182 km²). **Population:** 25,000,000. **Capital:** Algiers, 1,600,000. **Government:** One-party socialist republic. **Language:** Arabic; Berber; French. **Religion:** Islam. **Exports:** Oil and natural gas, citrus, wine, and dates. **Climate:** Mediterranean on the coast; hot and dry inland. □ After 130 years of colonial rule, France regarded Algeria as its southern province. When a national liberation movement began in the 1950s, France resisted it with unusual brutality. Independence was granted in 1962 and 1 million French settlers fled to Europe. Most of Africa's second largest nation (after Sudan) is nearly deserted. Only nomads will brave the Sahara (which covers 80% of Algeria); they include half a million Tuaregs. In their society, the men, not the women, wear veils. The Sahara contains Algeria's valuable oil and natural gas deposits. In the Great Western Erg, crescent-shaped sand dunes 400 ft. (122 m) high may drift 100 ft. (30 m) each year.

## CHAD B

**Area:** 495,800 sq. mi. (1,279,164 km²). **Population:** 5,450,000. **Capital:** N'Djamena, 250,000. **Government:** Military republic. **Language:** Arabic, many African dialects. **Religion:** Islam 45%, Christianity 30%, indigenous religions. **Exports:** Cotton and peanuts. **Climate:** Hot and dry in north. □ Landlocked and sparsely populated Chad is virtually two separate countries. The nomadic Muslim minority in the north has fought an intermittent 25-year war with the black African farmers in the South. In the late 1980s, French troops expelled Libyan forces aiding the Muslims. Lake Chad, which is also bordered by Niger, Nigeria, and Cameroon, is all that's left of a prehistoric inland sea. It contains fresh water, which is unusual for a desert lake. Though its depth and dimensions vary according to season, the lake continues to shrink.

## EGYPT C

**Area:** 386,660 sq. mi. (997,583 km²). **Population:** 53,000,000. **Capital:** Cairo, 6,300,000. **Government:** Republic. **Language:** Arabic; English. **Religion:** Islam. **Exports:** Cotton and cotton products, oranges, dates, oil, and rice. **Climate:** Hot and dry, with mild winters. □ Over 50 million people live in the narrow Nile valley, the world's largest oasis. Occupying only 4% of Egypt, this strip of land is 750 mi. (1,200 km) long and little more than 2-10 mi. (3.2-16 km) wide. The Nile delta fans out to include 155 mi. (250 km) of Mediterranean coastline. Cairo, the overcrowded desert capital, is Africa's largest city. Egypt is the world's leading producer of high-quality cotton, and after South Africa, it is the most industrialized country in Africa. But overpopulation and an explosive birthrate keeps it in poverty. Five thousand years ago, Egypt was one of the great civilizations. Each year over a million tourists come to see the pyramids, Africa's largest structures, which were built as tombs for the pharaohs. Ancient structures have been remarkably well preserved by the desert climate. For millenia, life in Egypt was sustained by the annual flooding of the Nile. When the Aswan High Dam was completed in 1968, greatly enlarging the capacity of an older dam, it became possible to control the flow of the river, store water for irrigation (doubling the agricultural output), and produce hydroelectric power. But without the flooding, no silt is deposited, farmers are forced to buy chemical fertilizers, and the delta region is shrinking. Before the Suez Canal was built in the 19th century, European and Asian shipping had to sail around the tip of Africa. There are plans to enlarge the canal because supertankers (loaded with Middle Eastern oil) must still take the longer route. Egypt is the political hub of the Arab world, and Al-Azhar University in Cairo is the center for Islamic learning. In 1979, after years of hostility, Egypt signed a peace treaty with Israel, which returned the Sinai Peninsula but still controls Egypt's Gaza Strip (Plate 27).

## LIBYA D

**Area:** 679,358 sq. mi. (1,752,744 km²). **Population:** 4,300,000. **Capital:** Tripoli, 900,000. **Government:** One-party socialist republic. **Language:** Arabic. **Religion:** Islam. **Exports:** Oil and oil products. **Climate:** The coast is Mediterranean; the rest is hot and dry. □ In 1959, oil was discovered under the Sahara, and Libya, previously one of the world's poorest nations, rapidly became the richest (per capita) in Africa. Except for two separate strips of fertile coastline, the nation is almost entirely a gravel-covered desert. From 1911 to World War II, it was a colony of Italy. After the war, Britain and France occupied Libya until it became an independent monarchy in 1951. Eighteen years later, a coup brought Colonel Muammar Qaddafi to power. He transformed Libya into a modern industrial state with strong educational and social welfare programs. In 1986, the US bombed Qaddafi's palace in Tripoli in retaliation for his support of international terrorism. This was not the first time the US had attacked the Libyan capital. In 1804, in response to harassment of shipping by pirates from the Barbary Coast (named after local Berber tribes), the US sent marines to the "shores of Tripoli" to destroy their base.

## MALI E

**Area:** 478,800 sq. mi. (1,221,042 km²). **Population:** 8,900,000. **Capital:** Bamako, 450,000. **Government:** Military republic. **Language:** French; African dialects. **Religion:** Islam 80%; indigenous religions. **Exports:** Cotton, peanuts, and dried fish. **Climate:** Desert region is hot; south is tropical. □ In the 14th century, the fabled city of Timbuktu ("from here to Timbuktu") was the center of Islamic learning and commerce in the vast North African kingdom of Mali (mah' lee). Today, it is a small desert city. In southern Mali, the Niger River forms an irrigation network of rivers and lakes (an "inland delta"). The largely black, Muslim population have faced drought, starvation, and disease.

## MAURITANIA F

**Area:** 398,000 sq. mi. (1,030,820 km²). **Population:** 2,100,000. **Capital:** Nouakchott, 170,000. **Government:** Military republic. **Language:** Arabic; French. **Religion:** Islam. **Exports:** Iron ore, copper, fish, gum arabic, and livestock. **Climate:** Hot and dry; the south is moderate. □ Mauritania is heavily dependent on aid from France. An encroaching Sahara covers 80% of the land, and desert sands are now blowing out over the Atlantic Ocean. The Senegal River, which separates Mauritania from Senegal, is the heart of the nation's only fertile region. Atlantic fishing and iron mining are the chief industries. Most Mauritanians are Moors, many of whom are nomads.

## MOROCCO G

**Area:** 172,415 sq. mi. (446,555 km²). **Population:** 24,000,000. **Capital:** Rabat, 525,500. **Government:** Constitutional monarchy. **Language:** Arabic; Berber; French. **Religion:** Islam. **Exports:** Phosphates, citrus, fish, produce, and crafts. **Climate:** Mediterranean on the coast; wide extremes in the mountains; hot in the desert. □ Eight miles (13 km) from Spain, across the Strait of Gibraltar, lies Morocco, "crossroads of Western and Islamic Culture." The old walled cities and native quarters within Casablanca and Marrakech are famous tourist attractions. Tangier, opposite Gibraltar, is an ancient trading port. The population is mostly mixed Arab and Berber. Arabic, French, and some Spanish are spoken in the cities; Berber is used in the Atlas Mountains. Morocco is the world's leading exporter of phosphates and is a major supplier of winter fruits and vegetables to Europe. Morocco was occupied by both Spain and France prior to independence in 1956. Spain retains control of Ceuta and Melilla on the Mediterranean coast.

## NIGER H

**Area:** 489,200 sq. mi. (1,267,028 km²). **Population:** 6,700,000. **Capital:** Niamey, 270,000. **Government:** Military republic. **Language:** French; African dialects. **Religion:** Islam. **Exports:** Uranium, peanuts, livestock, cotton, and fish. **Climate:** Hot and dry except in the south. □ Landlocked Niger (ni' jer) is named after a mighty river that flows only through a tiny portion of this often drought-stricken nation. Lake Chad, on its western border, is its other reliable source of water. Most of Niger's many ethnic tribes are black Muslims.

## SUDAN I

**Area:** 967,525 sq. mi. (2,505,890 km²). **Population:** 24,000,000. **Capital:** Khartoum, 550,000. **Government:** Military republic. **Language:** Arabic; English; African dialects. **Religion:** Islam 70%, indigenous 25%, Christianity 5%. **Exports:** Gum arabic, cotton, peanuts, sugar, and textiles. **Climate:** Hot and dry in the north; humid in the south. □ The largest nation in Africa is composed of two warring societies: Arabic-speaking Muslims in the north and blacks in the south. The latter have violently resisted attempts by the government in the north to impose Islamic law throughout the nation. The southern part of the country is the heart of Sudan's farming region; the best land lies between the White and Blue Nile Rivers. The three largest cities (Omdurman, Khartoum, and Khartoum North) are located where the rivers join to form the Nile. In the far south is the Sudd, one of the world's largest swamps. Sudan's acacia trees yield 90% of the world's gum arabic, a sticky substance used in adhesives, inks, candy, cosmetics, perfumes, and medicines. Sudan is struggling with war, drought, and the presence of thousands of refugees from Chad and Ethiopia. The refugees strain an already depleted agricultural sector, raising the specter of famine in the former "breadbasket of Africa."

## TUNISIA J

**Area:** 63,200 sq. mi. ( 163,688 km²). **Population:** 7,700,000. **Capital:** Tunis, 600,000. **Government:** Socialist republic. **Language:** Arabic; French. **Religion:** Islam. **Exports:** Oil, phosphates, olive oil, grapes, and dates. **Climate:** Mediterranean. □ Tunis, the capital city, is near the site of Carthage, a Phoenician city that dominated the Mediterranean for 1,000 years before being destroyed by the Romans in 146 BC. Ruins of the later Roman colony attract many tourists. After independence from France in 1956, Tunisia pursued a policy of moderation in foreign affairs under the skillful leadership of Habib Bourguiba. It was the most European of the north African nations, but after 30 years of Western secularism, there are signs of an Islamic revival.

## WESTERN SAHARA K

**Area:** 102,690 sq. mi. (265,967 km²). **Population:** 115,000. **Capital:** Aaiún, 34,000. **Government:** Controlled by Morocco. **Language:** Arabic; Berber. **Religion:** Islam. **Exports:** Phosphates. **Climate:** Mild and dry. □ When Spain gave up Spanish Sahara in 1975, Morocco received the phosphate-rich northern portion and Mauritania took the southern part. The Polisario, a national liberation organization aided by Algeria, resisted both countries. Mauritania gave up its portion in 1979 and Morocco claims all of Western Sahara as part of historic "Greater Morocco." The resistance continues.

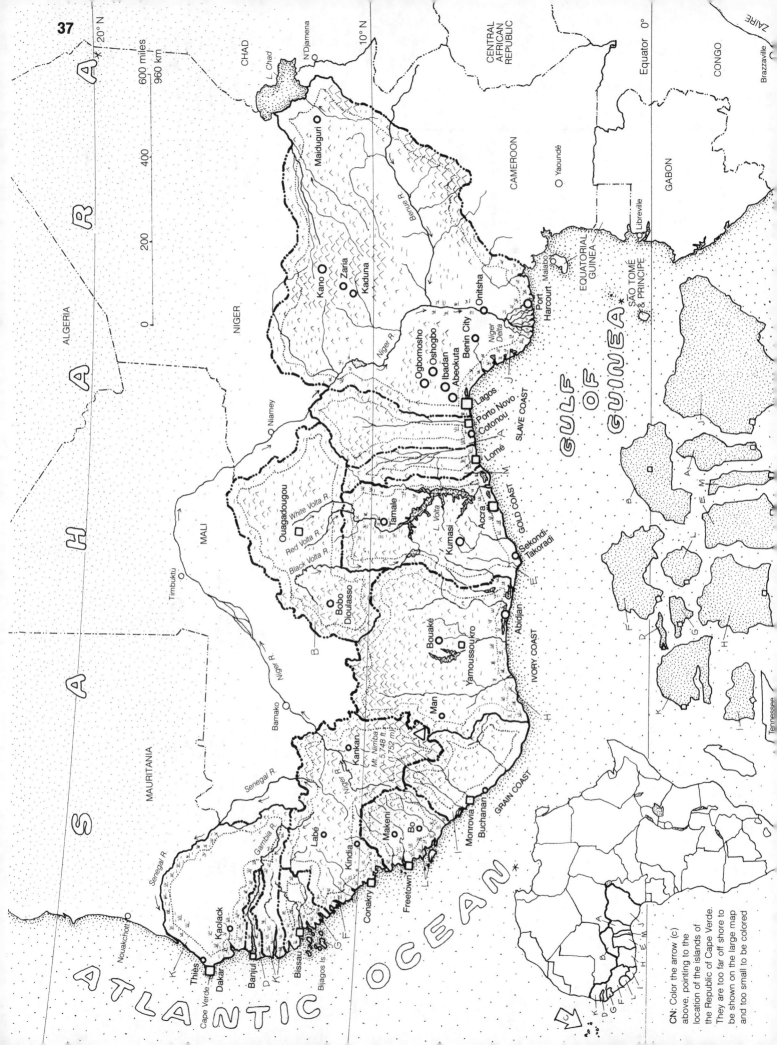

20° N

ALGERIA

10° N

CENTRAL
AFRICAN
REPUBLIC

Equator 0°

ZAIRE

S A H A R A

600 miles
960 km

400

200

0

MAURITANIA

MALI

NIGER

CHAD

N'Djamena

L. Chad

Maiduguri

Kano

Zaria

Kaduna

Benue R.

CAMEROON

Yaoundé

GABON

CONGO.

Brazzaville

Niger R.

Ogbomosho

Oshogbo

Ibadan
Abeokuta

Benin City

Onitsha

Niger
Delta

Port
Harcourt

Malabo

EQUATORIAL
GUINEA

Libreville

SÃO TOMÉ
& PRÍNCIPE

Niamey

Lagos

Porto Novo

Cotonou

Lomé

SLAVE COAST

GULF

OF

GUINEA

Timbuktu

Ouagadougou

White Volta R.

Red Volta R.

Black Volta R.

Tamale

L. Volta

Kumasi

Accra

GOLD COAST

Sekondi-
Takoradi

Senegal R.

Niger R.

Bamako

Bobo
Dioulasso

Bouaké

Yamoussoukro

Man

IVORY COAST

Abidjan

Senegal R.

Kankan

Mt. Nimba
5,748 ft.
(1,752 m)

Gambia R.

Labé

Kindia

Makeni

Bo

Monrovia

Buchanan

GRAIN COAST

Nouakchott

Thiès

Cape Verde

Dakar

Kaolack

Banjul

Bissau

Bijagos Is.

Conakry

Freetown

A T L A N T I C   O C E A N

Tennessee

# AFRICA: WESTERN

Many coastal nations on the western "bulge" of Africa share a similar landscape, climate, economy, religion, and history. Cash crops introduced by Europeans—cocoa, coffee, palm oil, and rubber—are the chief products of the coastal rain forests. Most of the world's cocoa and chocolate come from the seeds of local cacao plants. Peanuts (locally called groundnuts) and cotton are grown inland, on the higher and drier Sahel (the savanna bordering the Sahara). Harbors had to be built along the coast because shallow waters and treacherous surf prevented large ships from reaching shore. During colonial days, sections of the coast were named for their principal trade activities. Even though the commerce has changed, the descriptive labels remain: Grain Coast (Liberia); Ivory Coast (Ivory Coast); Gold Coast (Ghana); and Slave Coast (Togo, Benin, and Nigeria). From the Slave Coast and other ports on the continent, an estimated 10 million African slaves were sent to the New World between the 16th and the 18th centuries. Many did not survive the voyage. Slave ships transported three different cargoes in a triangular route: they carried finished goods to Africa to be exchanged for slaves who were shipped to the Americas for raw materials bound for Europe.

Christianity was introduced by European missionaries. Islam came from the sub-Saharan nations on the northern borders. But the majority of West Africans are animists—they worship the dead souls and spirits believed to be part of the natural environment.

When these nations gained independence (circa 1960), they retained either French, English, or Portuguese as official languages.

## BENIN A

**Area:** 43,480 sq. mi. (112,613 km²). **Population:** 4,500,000. **Capital:** Porto Novo, 140,000. **Government:** Military republic. **Language:** French; African dialects. **Religion:** Animism 70%; Christianity 15%; Islam 15%. **Exports:** Palm oil, cotton, cocoa, and peanuts. **Climate:** Hot, with coastal rainfall. ☐ Benin (beh neen'), formerly Dahomey, has had more coups than any of the other newly independent African nations. Most of its population practice animism (Benin is considered the birthplace of West Indian voodoo and black magic).

## BURKINA FASO B

**Area:** 105,875 sq. mi. (274,216 km²). **Population:** 8,750,000. **Capital:** Ouagadougou, 290,000. **Government:** Military republic. **Language:** French; African dialects. **Religion:** Animism 65%; Islam 25%; Christianity 10%. **Exports:** Livestock, peanuts, and cotton. **Climate:** Warm and generally dry. ☐ Burkina Faso, formerly Upper Volta, is an extremely poor landlocked country with few natural resources. Though the Black, Red, and White Volta Rivers flow through the country, it is vulnerable to droughts. The thin, nonporous soil is unable to absorb the summer monsoon rains that sweep up from the Gulf of Guinea. In addition to aid from France, the economy is helped by paychecks sent home by citizens working on the Guinea coast.

## CAPE VERDE C

**Area:** 1,557 sq. mi. (4,033 km²). **Population:** 340,000. **Capital:** Praia, 50,000. **Government:** Republic. **Language:** Portuguese; Crioulo. **Religion:** Roman Catholic. **Exports:** Coffee, bananas, salt, coal, and fish. **Climate:** Hot and very dry. ☐ About 300 mi. (480 km) off the coast of Senegal lie the islands of Cape Verde (verd). They were uninhabited when the Portuguese arrived in the 15th century. Plantations were built, and slaves were brought from the mainland. Cape Verde was also used as a training base for slaves bound for the Americas. Prolonged droughts have sent half the population away.

## GAMBIA D

**Area:** 4,360 sq. mi. (11,292 km²). **Population:** 800,000. **Capital:** Banjul, 45,000. **Government:** Republic. **Language:** English; African dialects. **Religion:** Islam. **Exports:** Peanuts. **Climate:** Tropical and humid. ☐ The smallest nation on continental Africa straddles the Gambia River. There are no bridges, so the river virtually creates two countries, each about 200 mi. (320 km) long and ten mi. (16 km) wide. The nation resembles a crooked finger poking deep into Senegal. Though the people of Gambia and Senegal share similar ethnic backgrounds and are allied in defense and foreign affairs, their dissimilar cultures (Gambia is English and Senegal is French) prevent unification. Gambia is where author Alex Haley went to trace his Roots.

## GHANA E

**Area:** 91,910 sq. mi. (238,047 km²). **Population:** 14,000,000. **Capital:** Accra, 800,000. **Government:** Military republic. **Language:** English; African dialects. **Religion:** Animists 45%; Christianity 40%. **Exports:** Cocoa, hardwoods, bauxite, diamonds, and gold. **Climate:** Tropical and wet. ☐ Ghana (gah' nuh) was named after a medieval African kingdom. Its former name, the Gold Coast, referred to a time when it was a gold-producing British colony. Ghana was once the world's leading exporter of cocoa which is still its major crop. Lake Volta, behind a dam on the Volta River, is one of the world's largest artificial lakes. The dam supplies power for a growing aluminum industry.

## GUINEA F

**Area:** 94,957 sq. mi. (245,939 km). **Population:** 6,200,000. **Capital:** Conakry, 550,000. **Government:** Military republic. **Language:** French; African dialects. **Religion:** Islam 70%; animism 25%; Christianity 5%. **Exports:** Bauxite, iron ore, bananas, coffee, and pineapples. **Climate:** Tropical, with heavy coastal rainfall. ☐ The word guinea (gihn' ee) is believed to be Berber, meaning "land of the blacks." Guinea, formerly French Guinea, was the first French colony to become independent (1958). Despite huge mineral reserves, Guinea's economy stagnated under years of communist rule. Guinea is the world's number two producer of bauxite (after Australia). The English "guinea" coin was minted from gold once mined here. Western Africa's three major rivers (the Niger, the Senegal, and the Gambia) originate in the plateau region.

## GUINEA-BISSAU G

**Area:** 13,948 sq. mi. (36,125 km). **Population:** 925,000. **Capital:** Bissau, 125,000. **Government:** Republic. **Language:** Portuguese; African dialects. **Religion:** Animism 65%; Islam 30%; Christianity 5%. **Exports:** Peanuts, coconuts, palm oil, and fish. **Climate:** Tropical and wet. ☐ Guinea-Bissau (bih sow'), formerly Portuguese Guinea, has the unhappy distinction of having lived under the longest period of colonial rule in history—over 500 years. It also suffered through 12 years of bitter fighting before it became Portugal's first African colony to achieve independence in 1974. Much of the country is a low-lying region and swampy. The offshore Bijago Archipelago is unusual—islands are rare on the coast of Africa.

## IVORY COAST H

**Area:** 124,500 sq. mi. (322,455 km). **Population:** 11,000,000. **Capital:** Yamoussoukro, 100,000. **Government:** One-party republic. **Language:** French; African dialects. **Religion:** Animism 65%; Islam 23%; Christianity 12%. **Exports:** Coffee, cocoa, hardwoods, bananas, rubber, and fish. **Climate:** Tropical, with coastal rainfall. ☐ The Ivory Coast was so named when there was an active trade in elephant tusks. It might better be called the "cocoa coast"—it is the world's top cocoa producer. It also one of the leaders in hardwood and coffee production. Félix Houphouet-Boigny, the only president since independence in 1960, has pursued a liberal economic policy which, until the commodity price decline of the 1980s, made the Ivory Coast the richest nation in black Africa. The booming port city of Abidjan (1,500,000), the "Paris of Africa," is the center of commerce and manufacturing. It was the capital until a new one was built inland at Yamoussoukro. There the president financed the construction of the world's largest Christian church.

## LIBERIA I

**Area:** 42,950 sq. mi. (111,241 km²). **Population:** 2,450,000. **Capital:** Monrovia, 450,000. **Government:** Republic. **Language:** English; African dialects. **Religion:** Animism 75%, Islam 15%; Christianity 10%. **Exports:** Iron ore, rubber, cocoa, coffee, and hardwoods. **Climate:** Tropical. ☐ Liberia became Africa's first independent black nation in 1847 (Haiti was the world's first). The land was purchased in 1822 by the American Colonization Society, an organization seeking to create an African homeland for freed American slaves. Descendants of those original immigrants make up less than 5% of the population, but they have traditionally controlled Liberia. These English-speaking "Americo-Liberians" are well-educated and prosperous; they live an American life-style. In 1980, a military coup sought to share that power among the many ethnic groups. Its leader, Samuel K. Doe, became president; but in 1990, he was killed and his corrupt government was overthrown in a vicious civil war. US interest in Liberia is probably the reason the nation was one of only two in Africa to escape colonization. Liberia is Africa's leading producer of iron ore and rubber. It has the world's largest merchant navy—foreign-owned ships use its registry because of low taxes.

## NIGERIA J

**Area:** 356,700 sq. mi. (923,853 km²). **Population:** 108,000,000. **Capital:** Lagos, 1,750,000. **Government:** Military republic. **Language:** English; African dialects. **Religion:** Islam 45%; Christianity 35%; animism 20%. **Exports:** Oil, cocoa, cotton, rubber, tin, palm oil, and peanuts. **Climate:** Tropical, with a dry interior. ☐ The most populous nation in Africa has the potential to become the continent's superpower. The land is rich in natural resources and capable of growing a wide variety of crops. The pyramids one sees in Nigeria are stacked sacks of peanuts awaiting shipment. Oil and natural gas produce 90% of the revenue that is financing industrial development. Nearly half of Nigerians are Muslims, who live mostly in the northern Sahel. The country tends to be divided along ethnic lines. The presence of over 250 different groups has led to numerous conflicts. In 1967, a bloody civil war was fought when the Ibos tried to create their own nation, Biafra, in eastern Nigeria.

## SENEGAL K

**Area:** 76,120 sq. mi. (197,151 km²). **Population:** 7,200,000. **Capital:** Dakar, 850,000. **Government:** Republic. **Language:** French; African dialects. **Religion:** Islam. **Exports:** Peanuts, cotton, phosphates, and fish. **Climate:** South coast is wet; interior is dry and hot. ☐ Until independence in 1960, Senegal (sehn ih gawl') was France's favorite part of French West Africa. Dakar, the "Gateway to Africa," was the colonial capital. Today it is a modern city; its magnificent harbor was the major supply port for Allied forces in Africa during World War II. Senegal's principal export is peanuts. The crop was originally brought to Africa by Europeans to feed the slave trade.

## SIERRA LEONE L

**Area:** 27,800 sq. mi. (72,002 km²). **Population:** 3,750,000. **Capital:** Freetown, 325,000. **Government:** Republic. **Language:** English; African dialects. **Religion:** Animism 75%; Islam; Christianity. **Exports:** Diamonds, bauxite, cocoa, and coffee. **Climate:** Tropical and wet. ☐ In 1787, Sierra Leone (lee own') was founded as the first colony for freed slaves by a British antislavery group. It remained a British colony for more than 150 years. After the British abolished slavery in 1807, they liberated slave ships and sent the freed slaves to Sierra Leone. The nation's main industry is the mining of diamonds, iron ore, and bauxite.

## TOGO M

**Area:** 21,650 sq. mi. (56,073 km²). **Population:** 3,400,000. **Capital:** Lomé, 235,000. **Government:** Republic. **Language:** French; African dialects. **Religion:** Animism 60%; Christianity 25%; Islam 15%. **Exports:** Phosphates, cacao, coffee, and palm oil. **Climate:** Tropical in the south. ☐ Tiny Togo is divided ethnically and physically. A mountain range separates northern Muslims, who are racially related to the people in Burkina Faso and Niger, from the southern Togolese who share ethnic backgrounds with tribes in Ghana and Benin. The country is poor, even with large phosphate deposits.

38

NIGER

Neb.
Colo. Kan. Mo.
N.M. Okl. Ark.
Texas La.

B

C

D

E

F

A

ANGOLA

B
E
F
D
A
H
A
C
I

L. Chad

N'Djamena

CHAD

SUDAN

10° N

NIGERIA

Niger R.

Benue R.

Garoua

B

Bouar

C

Mt. Cameroon
13,350 ft. (4,070 m)

Port Harcourt

Douala

Malabo

BIOKO

RÍO MUNI

Bata

PRÍNCIPE

São Tomé

Libreville

SAO TOMÉ

Port-Gentil

Lambarene

Yaoundé

Berbérati

Bangui

Ubangi R.

Mbandaka

Ogooué R.

Congo R. (Zaire R.)

Ubangi R.

Congo R. (Zaire R.)

Lualaba R.

Margherita Peak
16,762 ft. (5,109 m)

White Nile

UGANDA

L. Albert

Kampala

Equator 0°

L. Edward

L. Victoria

L. Kivu

Kigali

RWANDA

Bukavu

Bujumbura

BURUNDI

E

G

F

G

F

D

Stanley Pool

Brazzaville

Kinshasa

Kasai R.

Kananga

Mbuji-Mayi

MITUMBA MOUNTAINS

TANZANIA

L. Tanganyika

Pointe-Noire

CABINDA
(Angola)

A
H
A

Congo R. (Zaire R.)

Cuango R.

L. Mweru

MITUMBA MOUNTAINS

H

ATLANTIC

Luanda

L. Bangweulu

L. Nyasa
(L. Malawi)

10° S

OCEAN

Lubumbashi

MUCHINGA MOUNTAINS

MALAWI

Mt. Môco
8,596 ft. (2,620 m)

Kitwe

Ndola

Lilongwe

Lobito

Benguela

Huambo

Zambezi R.

Cuando R.

Cubango R.

Lusaka

Victoria Falls

L. Kariba

MOZAMBIQUE

Cunene R.

Zambezi R.

Harare

ZIMBABWE

A

0      200      400      600 miles

960 km

20° S

NAMIBIA

BOTSWANA

# AFRICA: CENTRAL

Dense rain forests form a wide band across central Africa (also called equatorial Africa). To Western explorers the impenetrable jungle, with its opaque canopy of vegetation, presented a dim, forbidding, and unknown world; they described Africa as "the Dark Continent." Few people live in the jungle. Most live to the north and south, up in the drier, tree-dotted savannas. Central Africa's sparse population is not just the result of an inhospitable jungle. Diseases, particularly malaria and sleeping sickness, take thousands of lives. Sleeping sickness is spread by the tsetse fly, whose bite can be just as deadly to certain domestic animals. The fly makes it virtually impossible to raise cattle in this part of Africa (along with many parts of western and eastern Africa). Because cattle are the primary source of fertilizer for most Third World nations, the tsetse fly also affects crop production.

Most Central Africans speak variations of the Bantu language. The Bantus displaced the pygmies some 2,500 years ago. The current pygmy population of 200,000 live as hunter-gatherers in remote parts of the jungle. They average 4½ ft. (1.4 m) in height. Pygmies are slowly giving up their jungle life-style, mostly because of the destruction of the rain forest.

The mighty Congo River (locally called the Zaire) and its hundreds of tributaries drain the world's second largest river basin (after the Amazon). The waterways serve as national highways but most of them are obstructed by rapids and waterfalls. In some places, mini-railroad lines transport cargo around river obstacles. Brazzaville and Kinshasa, the capital cities of the Congo and Zaire, face each other from opposite banks of the Congo River.

Portugal was the first European nation to explore and colonize lands that border central Africa's Atlantic coast. But here, as in West Africa, the Portuguese lost most of their possessions to more powerful European nations.

## ANGOLA A

**Area:** 481,360 sq. mi. (1,240,092 km²). **Population:** 9,700,000. **Capital:** Luanda, 500,000. **Government:** Republic (communist). **Language:** Portuguese; Bantu dialects. **Religion:** Christianity 70%, indigenous 20%. **Exports:** Oil, coffee, diamonds, and sisal. **Climate:** Generally mild. □ Any nation blessed with the rare combination of natural resources, accessible coastline, fertile land, and a favorable climate should be quite prosperous. But a 14-year war of independence, 14 more years of civil war, a communist-controlled economy, and the abrupt departure of the Portuguese in 1975, have left Angola in shambles. Only the flow of oil has been unaffected by the prolonged upheaval. The civil war showed signs of ending in 1989 when governnment forces (aided by Cuban troops and USSR equipment) and the UNITA guerrilla movement (supported by the US and South Africa) agreed to negotiate a settlement. Angola was the last Portuguese colony in Africa to gain independence. During the period of slavery, 2 million Angolans were sent to Brazil and other parts of Portugal's empire.

## CAMEROON B

**Area:** 183,569 sq. mi. (475,444 km²). **Population:** 11,500,000. **Capital:** Yaoundé, 325,000. **Government:** One-party republic. **Language:** French; English; African dialects. **Religion:** Indigenous 50%; Christian 35%; Islam 15%. **Exports:** Oil, natural gas, cocoa, coffee, timber, rubber, palm oil, and cotton. **Climate:** Hot and extremely wet on the coast; drier in the north. □ For a nation with over 200 ethnic tribes, Cameroon has had an usually stable government. Culturally, Cameroon reflects its multicolonial past. It was first ruled by Germany in the late 19th century, then divided between Britain and France after their victory in World War I. The British and French Cameroons united in 1960 (the northern part of British Cameroon joined Nigeria). Cameroon is the only country in Africa that uses both French and English as official languages. The climate and the mountainous landscape change dramatically from the arid north to the extremely wet and green coastal region. Each year, Mt. Cameroon (13,354 ft., 4,070 m), an active volcano and the tallest peak in this part of Africa, attracts 400 in. (1,016 cm) of rain, making it the wettest spot on the continent.

## CENTRAL AFRICAN REPUBLIC C

**Area:** 240,530 sq. mi. (622,947 km²). **Population:** 2,850,000. **Capital:** Bangui, 300,000. **Government:** Military republic. **Language:** French; African dialects. **Religion:** Indigenous 60%, Christianity 20%, Islam 10%. **Exports:** Diamonds, coffee, and cotton. **Climate:** Tropical; dry in the northeast. □ When it was part of French Equatorial Africa, this poor, landlocked nation was called Ubangi-Shari. The name derives from two rivers: the Chari, which flows north to Lake Chad, and the Ubangi. The Ubangi joins the Congo (Zaire) River in the south and provides the only feasible route to the sea: 1,000 miles (1,600 km) with a short rail connection to the coast. For three terrible years, 1976-1979, the nation was called the Central African Empire and its leader, Jean-Bedel Bokassa, declared himself emperor. His opulent inauguration made international headlines. Bokassa brutalized his people, and his personal trade in diamonds and ivory not only bankrupted the nation but nearly wiped out its elephant population.

## CONGO D

**Area:** 132,050 sq. mi. (342,010 km²). **Population:** 2,100,000. **Capital:** Brazzaville, 450,000. **Government:** Republic (communist). **Language:** French, Bantu dialects. **Religion:** Indigenous (50%), Christianity (40%). **Exports:** Oil, timber, wood products, potash, uranium, palm oil, and tobacco. **Climate:** Equatorial and wet. □ The Congo is *not* the former Belgian Congo, which is now Zaire. The Congo had the first communist government in Africa, but this did not prevent it from having the West pump its oil and

mine its minerals. Brazzaville was the former capital of French Equatorial Africa, which included Gabon, Chad, and the Central African Republic. Much of the Congo's export revenue has been squandered on poor planning and impractical projects. The nation is nearly covered by impenetrable jungle or nonarable soil; consequently, over half of the population live in cities and towns—an usually high percentage for Africa. This concentration of potential converts has aided the work of Christian missionaries.

## EQUATORIAL GUINEA E

**Area:** 10,830 sq. mi. (28,050 km²). **Population:** 365,000. **Capital:** Malabo, 37,000. **Government:** One-party republic. **Language:** Spanish, Bantu. **Religion:** Roman Catholic 80%, indigenous 20%. **Exports:** Cocoa, coffee, and timber. **Climate:** Equatorial and very wet. □ Equatorial Guinea is a tiny nation of two parts: an island and a larger mainland portion. Malabo, the capital, is on Bioko, an island 100 mi. (160 km) northwest of Rio Muni. (Bioko is only 20 mi. (32 km) from the coast of Cameroon.) Under Spanish rule, the island's fertile volcanic soil (with heavy rain) produced the world's finest cocoa. The colony had the highest per capita income in Africa. Equatorial Guinea is the only nation in Africa to use Spanish as its official language. The country is still recovering from the devastating rule of its first independent government—11 years of brutal rule and economic mismanagement.

## GABON F

**Area:** 103,340 sq. mi. (267,651 km²). **Population:** 1,500,000. **Capital:** Libreville, 360,000. **Government:** One-party republic. **Language:** French; Bantu dialects. **Religion:** Indigenous 50%; Christianity 50%. **Exports:** Oil, manganese, iron ore, hardwoods, and uranium. **Climate:** Equatorial and damp. □ Oil and the world's largest production of manganese have made thinly populated Gabon (gah bone') the wealthiest black nation in Africa. But little of the prosperity has filtered down to the people. At one time Gabon's income came solely from the sale of hardwoods. Tourists visit Lambarene to see the hospital built in 1913 by the young Albert Schweitzer. The great physician, musician, philosopher, and humanitarian spent 53 years ministering to the needs of the African people. Contrasted to that body of selfless work is the new $3 billion rail line that runs 400 mi. (640 km) from Libreville, the coastal capital, to Francoville, the birthplace of Gabon's president, Omar Bongo.

## SÃO TOMÉ & PRINCIPE G

**Area:** 371 sq.mi. (957 km²). **Population:** 106,000. **Capital:** São Tomé. **Government:** One-party republic. **Language:** Portuguese. **Religion:** Roman Catholic. **Exports:** Cocoa, coffee, bananas, palm oil, and copra. **Climate:** Equatorial. □ The tiny nation of São Tomé and Principe (soun' tuh may' & prin' suh pay), consists of two mountainous islands and two islets 200 mi. (320 km) west of Gabon, in the Gulf of Guinea. São Tomé has 90% of the nation's land and a similar percentage of the population. The islands were uninhabited when the Portuguese discovered them in 1470. They were first used as a penal colony. Blacks were brought from the mainland to work the sugar plantations. Working conditions hardly improved after slavery was abolished; under Portuguese management hundreds of workers were killed in labor protests. Much of the current work force comes from the former Portuguese colonies Cape Verde, Angola, and Mozambique. Cocoa is still the principal export, though production is less than it was when São Tomé, under Portuguese rule, was the world's leading producer.

## ZAIRE H

**Area:** 905,360 sq. mi. (2,335,828 km²). **Population:** 33,500,000. **Capital:** Kinshasa, 2,350,000. **Government:** One-party republic. **Language:** French; 200 Bantu dialects. **Religion:** Christianity 55%; indigenous 35%; Islam 10%. **Exports:** Cobalt, industrial diamonds, copper, oil, coffee, uranium, and palm oil. **Climate:** Equatorial, but seasonal in the south. □ Zaire (zah eer'), formerly the Belgian Congo, is a huge, very poor nation with enormous untapped mineral wealth and hydroelectric potential. Since independence in 1960 the nation has been beset by secessionist movements. The most serious challenges to national unity were the revolts of Katanga province (one of the world's richest mineral regions) from 1960-1963, and again in 1977. Under President Mobutu Sese Seko, who has held power since 1965, the country has been struggling with widespread corruption and ill-conceived economic projects. An additional obstacle to industrial development is the general lack of transportation facilities. Most of Zaire is an equatorial jungle basin drained by the Congo (Zaire) and hundreds of tributaries. These waterways form the nation's highways. On a trip up the Congo Joseph Conrad wrote his famous story "Heart of Darkness," in which he described the journey as a trip "to the earliest beginnings of the world."

## ZAMBIA I

**Area:** 290,000 sq. mi. (748,200 km²). **Population:** 7,500,000. **Capital:** Lusaka, 500,000. **Government:** One-party republic. **Language:** English; many African dialects. **Religion:** Christianity 60%, animism 40%. **Exports:** Copper and other metals. **Climate:** Tropical, but cooler at higher elevations. □ Mineral-rich but landlocked Zambia (formerly Northern Rhodesia) has struggled for years to get its exports to market, using the ports of neighboring Angola, Mozambique, and Tanzania. Railway lines through those nations have been the frequent target of guerrilla activity. In the 1980s the government was prompted to develop a broad agricultural base to offset the critical dependence on declining copper revenues (Zambia is the world's third largest copper producer). The landscape is a savanna-covered, relatively flat plateau with large swampy areas. The country was named after the Zambezi River, which forms the southern border with Zimbabwe. It was there that the English explorer David Livingstone discovered Victoria Falls which he named in honor of his queen. The falls present an awesome visual and auditory experience—its local name means "the smoke that thunders."

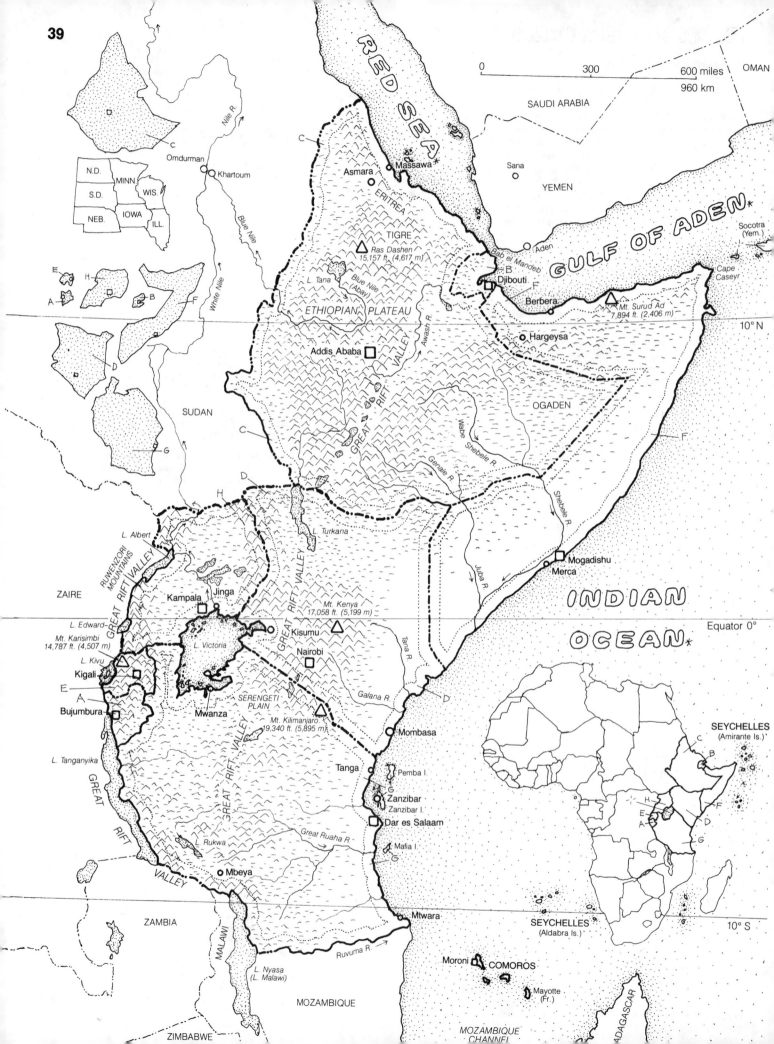

39

RED SEA

0   300   600 miles
          960 km

OMAN

SAUDI ARABIA

Sana

YEMEN

GULF OF ADEN

Socotra
(Yem.)

Cape
Caseyr

Aden

Bab el Mandeb

Massawa
Asmara

ERITREA

TIGRE

Ras Dashen
15,157 ft. (4,617 m)

L. Tana

Blue Nile
(Abay)

ETHIOPIAN PLATEAU

Addis Ababa

Djibouti

Berbera

Mt. Surud Ad
7,894 ft. (2,406 m)

Hargeysa

10° N

Awash R.

GREAT RIFT VALLEY

OGADEN

Wabe Shebele R.

Genale R.

Omdurman

Khartoum

Nile R.

Blue Nile

White Nile

N.D.   MINN.
S.D.   WIS.
NEB.  IOWA  ILL.

SUDAN

E   H
A       B
        F
    D
        G

C

C

D

H

L. Albert

RUWENZORI MOUNTAINS

GREAT RIFT VALLEY

L. Edward

Mt. Karisimbi
14,787 ft. (4,507 m)

L. Kivu

Kigali

E

A

Bujumbura

L. Tanganyika

ZAIRE

Kampala   Jinga

L. Turkana

Kisumu

Nairobi

Mt. Kenya
17,058 ft. (5,199 m)

L. Victoria

SERENGETI PLAIN

Mwanza

Mt. Kilimanjaro
19,340 ft. (5,895 m)

GREAT RIFT VALLEY

Tana R.

Galana R.

Shebele R.

Juba R.

Mogadishu

Merca

INDIAN OCEAN

Equator 0°

D

F

Mombasa

Tanga

Pemba I.

Zanzibar
Zanzibar I.

Dar es Salaam

Mafia I.

GREAT RIFT VALLEY

L. Rukwa

Great Ruaha R.

Mbeya

ZAMBIA

MALAWI

L. Nyasa
(L. Malawi)

Mtwara

10° S

Ruvuma R.

ZIMBABWE

MOZAMBIQUE

MOZAMBIQUE CHANNEL

Moroni   COMOROS

Mayotte
(Fr.)

MADAGASCAR

SEYCHELLES
(Amirante Is.)

SEYCHELLES
(Aldabra Is.)

C
B
H
E
A
F
D
G

# AFRICA: EASTERN

The conditions that created the Ethiopian famines of the 1980s still persist: limited rainfall, agricultural policies that discourage food production in favor of cash crops, the forced transfer of populations, and a continuing civil war. The dark-skinned and fine-featured people of this region are Caucasoids of Hamitic origin, related to the people of the Middle East. Christianity and Islam are the dominant religions. South of the Horn live mostly black Africans of the Swahili-speaking Bantu tribes. The nations of this region are considerably drier, higher, and cooler than other equatorial countries in central and western Africa. The Great Rift Valley is the major geological feature that separates eastern Africa from the rest of the continent (see Plate 35).

## BURUNDI A

**Area:** 10,747 sq. mi. (27,834 km²). **Population:** 5,000,000. **Capital:** Bujumbura, 170,000. **Government:** One-party republic. **Language:** Kirundi; French. **Religion:** Roman Catholic 70%; indigenous 25%. **Exports:** Coffee, cotton, and tea. **Climate:** Mild. ☐ About 85% of the people of Burundi (buh run' dee) population are Hutu farmers, but for three centuries they have been ruled by Tutsi (Watusi) cattle herders. The Tutsi are few in number but large in physical stature (a height of 7 ft. is common). They own whatever wealth there is in this extremely poor nation. In 1972, over 150,000 Hutus died in an unsuccessful rebellion. Thousands fled north to Rwanda, where a Hutu majority holds power over the Tutsi. Burundi was the southern part of Ruanda-Urundi, a Belgian Trust Territory, until independence was granted in 1962. This landlocked nation has no easy route to world markets for its mountain-grown coffee.

**COMOROS.** Over 400,000 people of mixed African, Middle Eastern, and South Asian ancestry live on three main islands (694 sq. mi., 1,794 km²) in the Mozambique Channel. Most are Muslims who speak Swahili, Arabic, or French. Comoros, dependent on French aid, lacks raw materials and fertile soil. When Comoros became independent in 1975, *Mayotte,* the fourth main island, chose to remain a French possession.

## DJIBOUTI B

**Area:** 8,900 sq. mi. (23,051 km²). **Population:** 390,000. **Capital:** Djibouti, 155,000. **Government:** Republic. **Language:** Arabic; French; Somali; Afar. **Religion:** Islam. **Exports:** Livestock and hides. **Climate:** Hot and dry. ☐ Djibouti (jih boo' tee) is a terribly hot desert nation strategically located on the strait of Bab el Mandeb between the Gulf of Aden and the Red Sea. Only 20 mi. (32 km) away is the Arabian Peninsula. Djibouti, the capital city and chief port, serves as a shipping terminal for Addis Ababa, the capital of Ethiopia, 400 mi. (640 km) inland. Djibouti earns money as an entrepôt for Ethiopian trade and Suez shipping. Other revenue comes from France—in the form of aid and from purchases made by a resident French garrison (whose presence keeps Djibouti from being swallowed up by Ethiopia or Somalia). Djibouti's Muslim population fall into two ethnic groups: the Afars (related to Ethiopians) in the north and the Somali-speaking Issas (related to Somalis) in the south.

## ETHIOPIA C

**Area:** 472,400 sq. mi. (1,223,516 km²). **Population:** 43,000,000. **Capital:** Addis Ababa, 1,425,000. **Government:** Marxist military. **Language:** Amharic; Galla; Sidama; Arabic. **Religion:** Coptic Christian 45%; Islam 45%; indigenous 10%. **Exports:** Coffee, oilseeds, hides, cotton, and sesame. **Climate:** Extremely hot on the coast; cooler in the interior. ☐ Ethiopia (formerly Abyssinia) is one of the world's oldest Christian nations. Until the 44-year reign of Emperor Haile Selassie was terminated by the current Marxist military government in 1974, there was an unbroken chain of kings and emperors dating back to Biblical times. Most Ethiopians are dark-skinned ("Ethiopia" is Greek for "land of sunburned faces"). The ethnically and linguistically diverse population is divided into two groups: the Semitic language-speaking Christian ruling class of the north and central regions and the Cushitic-language speaking Muslims of the south and southeast. The high, rugged plateaus on which they live are similarly divided by the Great Rift Valley. In the higher northern plateau, the Blue Nile begins at Lake Tana and winds its way through the world's largest gorge (longer and wider than the Grand Canyon) en route to the Nile in Sudan. The Blue Nile provides 90% of the water that flows through the Nile; Egypt is concerned that irrigation dams may be built in Ethiopia. Addis Ababa, the modern capital city, sits on an 8,000 ft. (2,439 m) plateau in the center of the country. Ethiopia had no coastline when the Eritrean region on the Red Sea was occupied by Italy from 1890 to 1952. The French helped build the railroad from Djibouti to Addis Ababa, and most Ethiopian exports are shipped from Djibouti. Coffee has been exported for so long that some believe the word comes from "Kaffa," a local region. The government has been fighting protracted wars with secessionists in Eritrea and Tigre.

## KENYA D

**Area:** 219,790 sq. mi. (637,391 km²). **Population:** 24,000,000. **Capital:** Nairobi, 850,000. **Government:** One-party republic. **Language:** Swahili; English; native dialects. **Religion:** Christian 70%; indigenous 25%; Islam 5%. **Exports:** Coffee, tea, pyrethum, cashews, sisal, and cotton. **Climate:** Hot and humid on coast; mild in the highlands. ☐ Beautiful white beaches, spectacular mountain scenery, a pleasant climate, and numerous wildlife parks and game preserves have made Kenya an outstanding attraction. Revenue from visitors exceeds the sale of coffee, the principal export. For years, mile-high Nairobi was famous as the African safari (Arabic for "trip") capital. Kenya's coast was first settled by Arabs 2,000 years ago. Mombasa (350,000), the second largest city and chief port, was an Arab colony. Many nations have controlled the coast, but the British colonized all of Kenya. Toward the end of their rule, they faced the fierce Mau Mau rebellion. In 1963, independence was granted. Most Kenyans speak Swahili, a Bantu tongue containing many Arabic and some Portuguese words. Kenya's earliest human history dates back 2 million years. Fossil bones of remote ancestors were discovered in the Great Rift Valley. Except for the fertile cooler highlands, most of Kenya consists of hot, arid plains; home to a wide variety of wildlife. Kenya has the world's highest birthrate (over 4% annually), and is losing the ability to feed itself.

## RWANDA E

**Area:** 10,170 sq. mi. (26,340 km²). **Population:** 6,600,000. **Capital:** Kigali, 165,000. **Government:** One-party republic. **Language:** Kinyarwanda; French. **Religion:** Roman Catholic 65%; animism 35%; **Exports:** Coffee, tea, tin, tungsten, and pyrethrum. **Climate:** Mild because of altitude. ☐ Rwanda (roo wahn' da) is Africa's most densely populated country. In 1959, the Hutus, who make up 90% of the population, overcame six centuries of rule by the monarchist Tutsi tribe. The bloody rebellion sent many Tutsi fleeing south to Burundi, where they still hold power over the large Hutu majority. Rwanda and Burundi, formerly Ruanda-Urundi, were part of Germany's East African Empire prior to World War I. The area became a Belgian Mandate after the war, and was split into two nations at independence in 1962. The Great Rift Valley, Lake Kivu, and tall mountains border Rwanda in the west. The mountains give way to sloping plateaus on which "Robusta" coffee, the essential ingredient in instant coffee, is the principal crop. Erosion of Rwanda's topsoil is imperiling the nation's huge population.

**SEYCHELLES.** This 90-island archipelago (170 sq. mi., 440 km²) in the Indian Ocean, 1,000 mi. (1,600 km) from the African mainland, is home to 69,000 residents of mixed African and European descent. Portugal discovered the islands in the 16th century; France created a colony 200 years later. In 1814, the Seychelles (say' shells) were given to England, which granted independence in 1976. Farming is limited because of the granite and coral composition of the islands. Cinnamon grows wild, and only here can one find trees producing double coconuts weighing as much as 50 lb. (22.7 kg).

## SOMALIA F

**Area:** 246,200 sq. mi. (637,658 km²). **Population:** 7,850,000. **Capital:** Mogadishu, 425,000. **Government:** Military republic. **Language:** Somali. **Religion:** Islam. Export: Livestock, hides, bananas, frankincense, and myrrh. **Climate:** Extremely hot and dry on the Aden coast; more moisture to the south. ☐ Somalia (so mah' lee uh or mahl' ya), occupying the tip of the Horn of Africa, is a poor, hot, and arid nation of nomads. The only arable land, irrigated by two nonnavigable rivers, is in the south. By African standards, the population of Somalia is remarkably uniform. The Somali-speaking Muslims are distinguishable from each other only by which of four clans they belong to. Despite a 2,000-year oral tradition, the Somali language was unwritten until a system was devised in the 1970s. Prior to independence in 1960, the nation was divided into British Somaliland on the Gulf of Aden and Italian Somaliland on the Indian Ocean. Somalia has angered neighboring Ethiopia, Djibouti, and Kenya by encouraging Somali-speaking Muslims in those countries to secede and join Somalia. In 1977, Somalia invaded Ethiopia in order to annex the southern region of Ogaden. When the Soviet Union failed to support the attack (and actually helped Ethiopia repel the invasion), Somalia broke relations with the USSR and sought US military aid.

## TANZANIA G

**Area:** 364,890 sq. mi. (945,065 km²). **Population:** 22,700,000. **Capital:** Dar es Salaam, 800,000. **Government:** One-party socialist republic. **Language:** Swahili; English. **Religion:** Animism 40%; Christianity 30%; Islam 30%. **Exports:** Sisal, coffee, cotton, cloves, coconuts, and tobacco. **Climate:** Islands and coast are tropical; interior is mild. ☐ Tanzania (tan za nee' a) was created in 1964 when newly independent Tanganyika and Zanzibar united. The name "Zanzibar" refers to the group of offshore islands, to the largest island itself, and to its capital city. Zanzibar is the world's leading producer of cloves. In the early 19th century, the city of Zanzibar was an Arab sultanate and the major slave trading center for east Africa. Tanganyika was a part of German East Africa; it became a British protectorate after World War I. At first, Tanzania's socialism improved the quality of life, but its economy collapsed in the 1970's when oil prices rose and commodity prices fell. It now depends on massive Western aid. The capital city, Dar es Salaam, handles commerce for the landlocked nations to the west. For many years it has been Africa's most important Indian Ocean port. Tanzania's many natural wonders include snow-capped Kilimanjaro, Africa's tallest peak (19,340 ft., 5,895 m); Lake Tanganyika, one of the world's longest and deepest lakes, and the one with the most species of fish; Lake Victoria, the world's second largest freshwater lake (after Lake Superior); Olduvai Gorge, where fossils of some of the earliest human ancestors have been found; Selous, the world's largest game park; and Ngorongoro, the world's second largest volcanic crater—12 mi. (19 km) across—whose watered grass-covered floor is home to 30,000 animals.

## UGANDA H

**Area:** 91,140 sq. mi. (236,053 km²). **Population:** 15,500,000. **Capital:** Kampala, 465,000. **Government:** Republic. **Language:** Swahili; English; African dialects. **Religion:** Christianity: 60%; animism 25%; Islam 15%. **Exports:** Robusta coffee, tea, cotton, and copper. **Climate:** Mild, with adequate rainfall. ☐ Although landlocked, spectacularly beautiful Uganda (yoo gan' da)—the "pearl of Africa"—should be a prosperous nation. It has fertile land, a pleasant climate, ample rainfall, hydroelectric power, mineral deposits, and a direct rail link to the port of Mombasa in Kenya. But since 1972, beginning with the murderous 7-year rule of General Idi Amin (300,000 Ugandans died), the nation has been torn apart and the economy shattered by numerous coups, invasions, civil wars, and tribal conflict among its many ethnic groups. Over 15% of Uganda is covered by fish-stocked lakes and rivers. On the shores of Lake Victoria and close to Kampala, the modern capital, is the Entebbe airport, the site of a famous raid by Israeli commandos who freed a planeload of hostages.

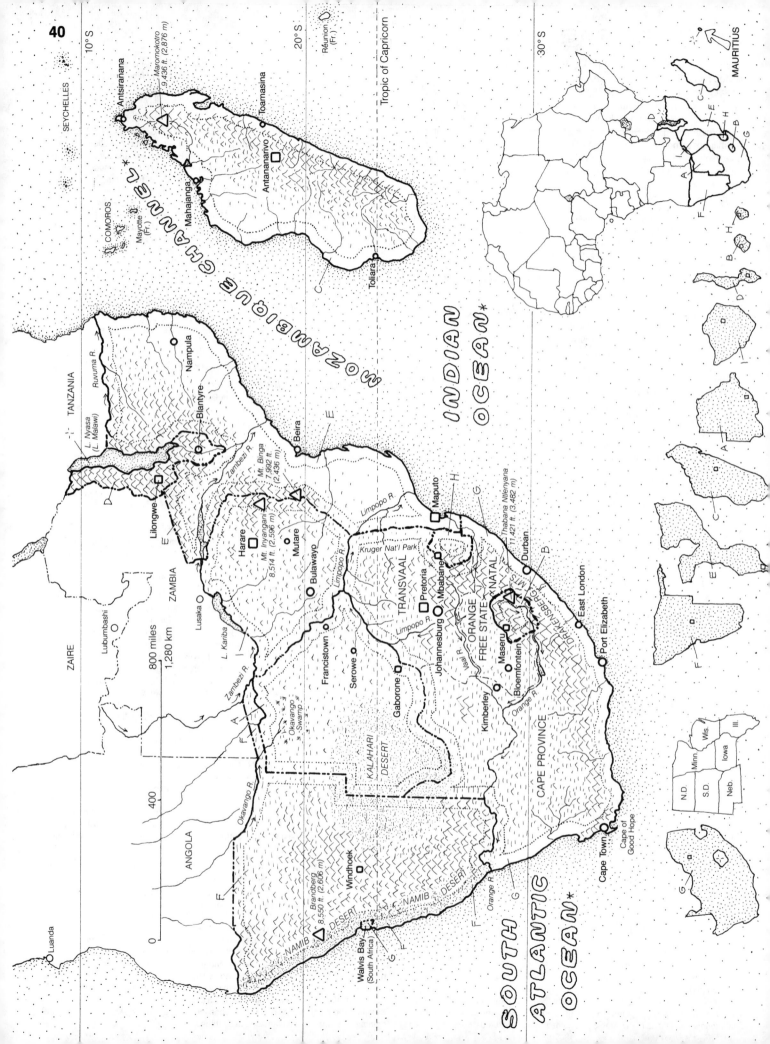

40

# AFRICA: SOUTHERN J

Southern Africa is mineral-rich. Its great natural wealth and good weather attracted the largest concentration of European immigrants in Africa. As a result, it has been most difficult for black majorities to regain control of this part of Africa. They are still powerless in South Africa, the dominant nation in the area. Most of southern Africa is a high plateau, sloping upward to the Drakensberg Mountains, which loom over the narrow coast. The western portion is covered by desert (the Namib) and semidesert (the Kalahari).

## BOTSWANA A

**Area:** 224,000 sq. mi. (580,160 km²). **Population:** 1,250,000. **Capital:** Gaborone, 65,000. **Government:** Republic. **Language:** English; Setswana. **Religion:** Indigenous 85%; Christianity 15%. **Climate:** Mild, with limited rainfall. ☐ The discovery of huge diamond and other metal deposits has brought prosperity to Botswana (bo tswa' na). The small population lives in the greener region along the border with South Africa. Though opposed to apartheid, Botswana depends on South Africa's business services, jobs, and rail links to the coast. Botswana was Bechuanaland, a British protectorate, until independence in 1966. The government is one of Africa's rare nonracial, multiparty republics. The land is a flat plateau, much of which is covered by the Kalahari Desert, home to the Bushmen, one of Africa's oldest races. The ancestors of these small yellowish-brown people retreated to the semidesert centuries ago. Their language is distinctive for its use of clicking sounds.

## LESOTHO B

**Area:** 11,720 sq. mi. (30,356 km²). **Population:** 1,700,000. **Capital:** Maseru, 50,000. **Government:** Military. **Language:** Sesotho; English. **Religion:** Christianity 80%; indigenous 20%. Exports: Wool, meat, diamonds. **Climate:** Temperate. ☐ Lesotho's (luh so' to) principal income consists of the paychecks sent home by the half of its male population who work in South Africa. Until 1966, this small nation was the British protectorate Basutoland, which kept it from becoming part of surrounding South Africa. Wealth in this mountainous country is measured by the ownership of livestock.

## MADAGASCAR C

**Area:** 226,650 sq. mi. (587,024 km²). **Population:** 11,100,000. **Capital:** Antananarivo, 600,000. **Government:** One-party republic. **Language:** Malagasy; French. **Religion:** Indigenous 55%; Christianity 40%; Islam 5%. Exports: Coffee, vanilla, spices, meat, and fish products. **Climate:** Tropical, with cool highlands and a wet east coast. ☐ Madagascar, 300 mi. (480 km) off the coast of Africa, is the world's fourth largest island. Millions of years ago, it was part of the mainland (Plate 2). Madagascar was originally settled by Indonesians and Malayans from Southeast Asia nearly 2,000 years ago. It was colonized by France in 1896 and received independence in 1960. Most Madagascans are of African and Asian descent. An ethnic division exists between the races. The black majority holds power, but the Indonesian influence has made Madagascar culturally closer to Asia. About 90% of the island's flora and fauna are unique to the island. The best-known animal is the monkeylike lemur (there are 22 species). Eggs of the prehistoric elephant bird are still being found; some weigh as much as 20 lb. (9 kg). The flightless creatures were killed off by early settlers, who began a process of extinction that is accelerating with the destruction of the eastern rain forest. The western part of the island is a deciduous woodland. Madagascar is in danger of losing its amazing variety of tree species.

## MALAWI D

**Area:** 45,747 sq. mi. (118,485 km²). **Population:** 7,000,000. **Capital:** Lilongwe, 99,000. **Government:** One-party republic. **Language:** Chichewa; English. **Religion:** Christianity 80%, Islam 20%. Exports: Tea, tobacco, peanuts, and cotton. **Climate:** Mild. ☐ The beautiful nation of Malawi (ma la' wee), was formerly Nyasaland, a British protectorate. The name "Malawi" comes from an early Bantu kingdom. Lake Nyasa (Lake Malawi) fills the southernmost trench of the Great Rift Valley. Malawi has had only one leader since independence in 1962: Dr. Hastings Kamuzu Banda, who made himself president for life. Unusually conservative for a black African leader, he promotes free enterprise, encourages foreign ownership of farmland, and maintains close relations with South Africa (where many Malawians work the mines). He has passed restrictive codes on personal appearance (long hair on men is forbidden, and women cannot wear shorts, pants, short skirts, and so on).

**MAURITIUS.** This group of tropical islands in the Indian Ocean, 500 mi. (800 km) east of Madagascar (marked by an arrow to the right of the small map of Africa) has been colonized by the Dutch, French, and British. The main island, Mauritius (maw rish' us) (790 sq.mi., 2,046 km²) has about 1 million multiracial citizens. Two-thirds are Hindus. Sugar plantations cover half of the land but tourism, fishing, and clothing manufacturing are growing industries.

## MOZAMBIQUE E

**Area:** 303,000 sq. mi. (784,770 km²). **Population:** 15,400,000. **Capital:** Maputo, 750,000. **Government:** One-party republic. **Language:** Portuguese, many native dialects. **Religion:** Animism 65%; Christianity 20%; Islam 15%. Exports: Cashews, tea, textiles, copra, and cotton. **Climate:** Tropical and humid along the coast. ☐ Mozambique (mo zuhm' beek) has struggled since gaining independence in 1975 (after 10 years of bloody fighting). When 250,000 Portuguese fled the country in fear of retaliation for centuries of brutality, they abandoned skilled positions that Mozambique citizens were unprepared to handle. The 1980s then brought both drought and deluge. The communist government made progress in education, health, and women's rights, but failed to improve the economy (it has since renounced Marxism). The greatest obstacle to national survival has been the barbaric war waged by right-wing rebels, organized and trained by white Rhodesians (before Rhodesia became Zambia and Zimbabwe) and South Africans, who for years had pursued a policy of destabilizing neighboring black-controlled governments. The guerrillas no longer receive outside aid, but they continue to terrorize the civilian population. They also interfere with the use of Mozambique railroads and shipping ports by the landlocked neighboring countries.

## NAMIBIA F

**Area:** 318,260 sq. mi. (824,293 km²). **Population:** 1,300,000. **Capital:** Windhoek, 115,000. **Government:** Republic. **Language:** English; Afrikaans; native dialects. **Religion:** Christianity 90%; indigenous 10%. Exports: Diamonds, copper, uranium, lead, and fish products. **Climate:** Temperate and very dry. ☐ Africa's last colony, Namibia (nuh mib' ee a), formerly South-West Africa, became independent in 1990 after 100 years of German and South African rule. South Africa ignored international criticism for 70 years by treating Namibia as a province. But mounting diplomatic pressure and the militant actions of the South-West Africa's People's Organization (SWAPO) persuaded South Africa to give up Namibia. The nation has a little more than 1 million people, including 70,000 whites from South Africa and about 30,000 Bushmen living in the Kalahari Desert. Most of Namibia is dry, particularly the Namib Desert along the Atlantic coast. The cold waters of Antarctica's Benguela Current (plate 44) are responsible for creating a coastal desert similar to those in Peru and Chile (Plate 17). Namibia's is believed to have the continent's largest diamond and uranium deposits.

## SOUTH AFRICA G

**Area:** 471,445 sq. mi. (1,235,304 km²). **Population:** 34,000,000. **Capital:** Pretoria, 450,000. **Government:** Republic. **Language:** Afrikaans; English; native dialects. **Religion:** Christianity 70%; indigenous 30%. Exports: Gold, coal, diamonds, food products, asbestos, and metal ores. **Climate:** Temperate and dry; Mediterranean on Cape coast. ☐ South Africa is the most industrial nation on the continent. It mines a wide array of minerals (some vital to US military needs) and half the world's supply of gold and diamonds. South Africa has been a strategic ally of the West, but in the 1980s, the US, western Europe, and other nations imposed economic sanctions (restrictions on trade and investment) against it because of its racial policies. Since 1948, the practice of "apartheid" (uh part' hite), based on skin color, has rigidly separated the population into four classes: blacks (who make up 70% of the population); whites (17%); colored (10%); and Asian (3%). Blacks have been divided ethnically into 10 separate "homelands" where they must live if not employed elsewhere. These homelands, generally located in unproductive parts of the country, occupy less than 15% of the land. The whites in South Africa have nearly segregated themselves as well. The white Afrikaners, the landowning majority, are concentrated in the interior regions of the Orange Free State and the Transvaal. These descendants of early Dutch, German, and French settlers speak Afrikaans, a dialect derived from Dutch and French. The Afrikaners control the military; their ruling Nationalist party created apartheid. The white English minority (40%) lives mostly in the Cape and Natal Provinces. These descendants of English colonialists run their own schools and run many of the nation's business and industrial communities. All whites in South Africa are taught both English and Afrikaans. The coloreds, who provide a skilled labor force in Cape Province, are mostly descendants of the earliest residents (Hottentots) and European settlers. The Asians living in Natal are descendants of contract laborers brought from India in the 19th century. The first Europeans to settle South Africa were the 17th century Dutch, called Boers ("farmers"). When the British arrived in the early 19th century, the fiercely independent Boers withdrew to the interior. Gold and diamonds were discovered in the interior; the British followed, resulting in the Boer Wars of 1880 and 1899-1902. The victorious British colonized all of South Africa, which became a self-governing nation in 1910. Since the 1970s, in response to black protests, international criticism, and economic sanctions, the government has been easing its racial restrictions. In 1990, it recognized the African National Congress (ANC), a militant black resistance organization, and released its leader, Nelson Mandela, after 27 years of imprisonment. The ANC renounced the use of violence and the government, under the leadership of President F.W. de Klerk, is moving toward the elimination of apartheid.

## SWAZILAND H

**Area:** 6,706 sq. mi. (1,751 km²). **Population:** 710,000. **Capital:** Mbabane, 33,500. **Government:** Monarchy. **Language:** siSwati, English. **Religion:** Christianity: 70%; indigenous 30%. Exports: Sugar, wood pulp, iron ore, asbestos, and citrus. **Climate:** Mild; cooler in highlands. ☐ Swaziland is one of Africa's three remaining kingdoms. It was a British protectorate until 1968, so it was not absorbed by South Africa, which surrounds it on three sides. The nation has good land, climate, water, and mineral deposits. Descendants of white immigrants from South Africa own nearly half the major farms, the mines, the industries, and the forests they planted in the highlands. Swaziland's vacation resorts operate as unrestricted playgrounds for South African whites.

## ZIMBABWE I

**Area:** 150,750 sq. mi. (390,452 km²). **Population:** 9,100,000. **Capital:** Harare, 675,000. **Government:** Republic. **Language:** Shona; Ndebele; English. **Religion:** Christianity 60%; indigenous 40%. Exports: Asbestos, chromium, gold, nickel, tobacco, and food products. **Climate:** Mild; ample rainfall. ☐ Zimbabwe (zim bah' bway) is named after an early African kingdom, whose stone ruins are a major tourist attraction. Most of the land is a fertile, high plateau with ample water, mineral deposits, and hydroelectric power. Though landlocked, it has rail connections to ports in Mozambique and South Africa. Zimbabwe was a British colony (Southern Rhodesia) until ruling whites defied Britain and asserted their independence in 1965. Fourteen years later, they yielded to international sanctions and guerrilla pressure and transferred power to the 98% black majority. Many whites were persuaded to stay on in key positions. About 4,000 prospering white farmers still own about a third of the country. Land reform has failed but educational programs have been too successful—there are far more qualified citizens than there are jobs.

# 41

# THE ARCTIC

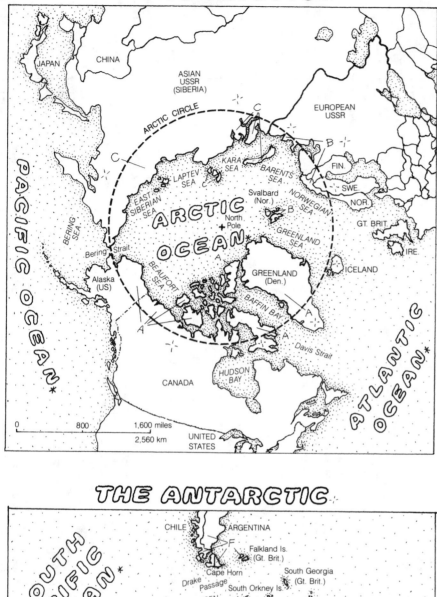

JAPAN
CHINA
ASIAN USSR (SIBERIA)
ARCTIC CIRCLE
EUROPEAN USSR
C
B
KARA SEA
BARENTS SEA
FIN.
LAPTEV SEA
C
Svalbard (Nor.)
NORWEGIAN SEA
SWE.
EAST SIBERIAN SEA
B
NOR.
ARCTIC OCEAN
GREENLAND SEA
GT. BRIT.
North Pole
A
IRE.
BERING SEA
Bering Strait
BEAUFORT SEA
GREENLAND (Den.)
ICELAND
Alaska (US)
BAFFIN BAY
A'
A
PACIFIC OCEAN *
Davis Strait
ATLANTIC OCEAN *
HUDSON BAY
CANADA
0    800    1,600 miles
2,560 km
UNITED STATES

## THE ARCTIC
### NORTH AMERICA A
### EUROPE B
### ASIA C

Arctic Circle
C
B
Arctic Ocean
Pacific Ocean
Atlantic Ocean

# THE ANTARCTIC

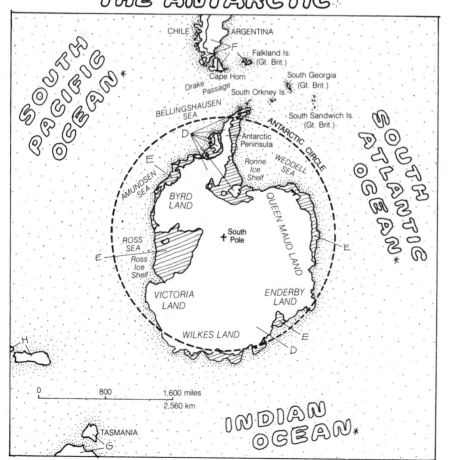

CHILE
ARGENTINA
F
Falkland Is. (Gt. Brit.)
Cape Horn
South Georgia (Gt. Brit.)
Drake Passage
South Orkney Is.
BELLINGSHAUSEN SEA
South Sandwich Is. (Gt. Brit.)
SOUTH PACIFIC OCEAN *
D
Antarctic Peninsula
ANTARCTIC CIRCLE
E
Ronne Ice Shelf
WEDDELL SEA
SOUTH ATLANTIC OCEAN *
AMUNDSEN SEA
BYRD LAND
QUEEN MAUD LAND
ROSS SEA
South Pole
E
Ross Ice Shelf
E
VICTORIA LAND
ENDERBY LAND
WILKES LAND
E
D
0    800    1,600 miles
2,560 km
H
TASMANIA
G
INDIAN OCEAN *

## THE ANTARCTIC
### ANTARCTICA D
### ICE SHELF E
### SOUTH AMERICA F
### AUSTRALIA G
### NEW ZEALAND H
### AFRICA I

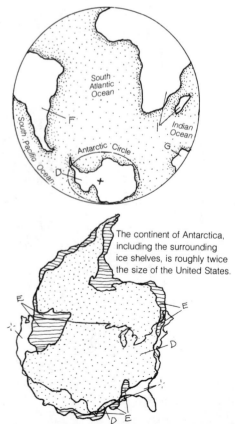

South Atlantic Ocean
F
Indian Ocean
South Pacific Ocean
G
Antarctic Circle
D

The continent of Antarctica, including the surrounding ice shelves, is roughly twice the size of the United States.

E
E
D
D E

# POLAR REGIONS

The polar regions are the lands bordered by the arctic circle in the northern hemisphere and the antarctic circle in the south. The circles mark the point at which the sun stays above the horizon for at least one 24-hour period during the year. Many geographers and scientists prefer to define the arctic as the region north of the "tree line" (the northernmost point at which trees will grow) or the "temperature line" (the line above which the average July temperature stays below 50° F, 10° C). Both lines roughly correspond to the astronomically determined arctic circle.

Because of limited precipitation, the polar regions can be classified as deserts. Antarctica is the driest, with an average annual snowfall equal to 2 in. (10 cm) of rain. Because of the cold, even small amounts of precipitation do not melt and are added to the ice cap. The arctic receives somewhat more moisture (6–10 in. or 15–25 cm) and because of poor drainage and slow evaporation, the arctic landscape remains marshy during the summer thaw.

Though the arctic and antarctic are comparable in size, have similar light and dark seasons, and receive very little precipitation, the two regions are in fact very different. The arctic is basically an ocean surrounded by the northern portions of North America, Europe, and Asia. The Arctic Ocean, the world's smallest, covers an area of 5,500,000 sq. mi. (14,374,500 km²). The antarctic region is just the opposite: it is itself a continent (Antarctica) surrounded by oceans (the South Atlantic, South Pacific, and Indian Oceans).

Airliners taking the shorter "great circle routes" routinely fly over the arctic, and nuclear submarines navigate under the ice at the north pole. Yet it wasn't until 1909 that explorers reached the "the top of the world." Admiral Robert E. Peary was the first man to set foot on the north pole. Two years later the Danish explorer Roald Amundsen was the victor by 34 days in the race to the south pole against a British team led by Robert F. Scott.

Because of warming ocean currents, the arctic is not quite as cold as the interior of Siberia. But much of the Arctic Ocean stays covered by 10–15 ft. (3–4.5 m) of ice. The relative warmth of summer causes part of the ice to break into moving packs, and the tundra (treeless, permanently frozen land) thaws out enough to support colorful plant growth. The thaw involves the uppermost 6 in. (15 cm) of the 1,000 ft. (305 m) or more of permafrost (permanently frozen ground).

The antarctic has no such dramatic change of season; it is on average 35° F (20° C) colder than the arctic. Gale-force winds combine with the frigid cold to produce the earth's fiercest weather. Winds in excess of 200 mph (320 kph) have been clocked, and the lowest temperature reading on record, –128° F (–89° C), was made during an antarctic winter. Floating ice shelves attached to various parts of the antarctic coastline considerably expand the size of the continent. In some places, the ice covering Antarctica is close to 3 mi. (4.8 km) thick. This ice cover contains 90% of the world's supply of fresh water. Most of Greenland (in the arctic) is very similar to Antarctica because both are under ice sheets of similar thickness and icebergs break off the glaciers on both land masses to form hazards to shipping.

Over 1 million people live within the arctic circle. Most are of Mongoloid ancestry, including the Inuit (Eskimos) of North America, the Lapps of Scandinavia, and the Chukchi and Samoyeds of the USSR. Because of the introduction of modern communication, transportation, scientific and mining operations into these regions, the life-styles of the natives have changed considerably. Snowmobiles are replacing dogsleds. The arctic is sparsely populated, but it seems crowded when compared to Antarctica, which hasn't a single permanent resident. About 4,000 scientists from many nations work there, mostly during the summer months.

The world's most vacant continent is also its highest: Antarctica's elevation averages 6,000 ft. (1,830 m). Some mountain ranges are as tall as 15,000 ft. (4,570 m) and the continent would be even higher if not for the weight of the thousands of feet of ice cover, which compresses land and mountains alike. The Antarctic Peninsula is an extension of the Andes Mountains of South America, less than 600 mi. (960 km) to the north.

The arctic has a rich sampling of land animals: polar bears, reindeer (in Europe and Asia), caribou (the reindeer's cousins in North America), wolves, foxes, and numerous smaller creatures. Except for polar bears, seals, walruses, and some foxes, most of the animals migrate south during the winter. Antarctica, on the other hand, is even more devoid of land animals than it is of humans. The only land animal present is a tiny, wingless mosquito, about a tenth of an inch (2.5 mm) long. Bird and sea life are abundant in both regions; Antarctica is best known for its penguin population. The Emperor penguin, the largest species, stands 4 ft. (1.2 m) tall and is capable of surviving the brutal antarctic winter.

Personnel in scientic stations and field laboratories are actively engaged in research throughout the polar lands. Meteorologists from both hemispheres gather data to assist in global weather forecasting. A variety of mineral deposits have been discovered, but mining operations, such as oil drilling on the north slopes of Alaska, are in progress only in the more accessible arctic.

Scientists are especially concerned about the status of the ozone layer of the earth's atmosphere. This barrier against ultraviolet radiation has been thinning rapidly because of certain contaminants released into the atmosphere by the industrialized nations. A hole has been discovered in the ozone layer above the south pole, and there has been a decrease in the region's phytoplankton, the plant source that feeds the shrimplike krill at the heart of the antarctic's marine food chain.

Though there are no immediate prospects for the exploitation of Antarctica, 16 nations have established permanent bases there, and 7 have staked out claims to ownership of the land. The United States and the rest of the world do not recognize these claims. The Antarctic Treaty, signed in 1959, grants nations the right to pursue scientific investigations for peaceful purposes, if they agree to share all discoveries. Military activity, nuclear testing, and the dumping of toxic waste are prohibited. An unfortunate amount of dumping and burning of waste and numerous fuel spills have already occurred, but efforts are being made to eliminate those practices. Nations are debating the merits of a new treaty, the Wellington Convention, that would allow mining in the antarctic (under the strictest supervision). Skeptical environmentalists fear the unbridled despoliation of the only remaining truly wild continent; along with a minority of nations, they favor a proposal that would make Antarctica a "world park."

## SUN, SUNLIGHT, SEASONS AT THE POLES.

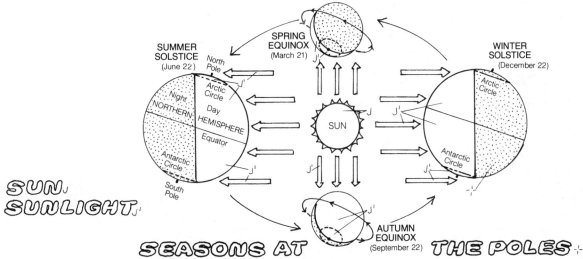

The earth revolves around the sun once every 365¼ days. The earth also revolves on its own axis (an imaginary line connecting the two poles) once every 24 hours, creating a night and day effect for most points on the planet. The earth's axis is tilted at an angle of 23.5° to the plane of its orbit around the sun. We have seasonal changes on earth because of this tilt. Without the tilt, the same amount of sunlight would fall on a particular point on the planet every day, and climates would not vary throughout the year. At the summer solstice, around June 22, the northern hemisphere is tilted toward the sun and experiences its longest days and shortest nights. The diagram shows that on this day the entire region within the arctic circle receives 24 hours of sunshine. Simultaneously, the antarctic region is in 24 hours of darkness, and winter is beginning in the southern hemisphere.

At the winter solstice, on December 22, the northern hemisphere is tilted farthest from the sun, and the entire arctic circle is in darkness for the first 24 hours of winter. The spring and autumn equinoxes occur around March 21 and September 22. The equinoxes fall midway between the solstices. On the days of the equinoxes, the earth's axis is perpendicular to the sun's rays—the axis is neither toward nor away from the sun (it might help to visualize the tilt as being "sideways" to the sun). The sun is directly above the equator, and days and nights are of equal length at all latitudes, in both hemispheres. Between the spring and summer equinoxes, regions within the arctic circle will experience at least one day of 24-hour sunlight ("midnight sun"), up to a maximum of 6 months at the north pole. During that period, an equivalent amount of darkness will prevail within the antarctic circle.

**42**

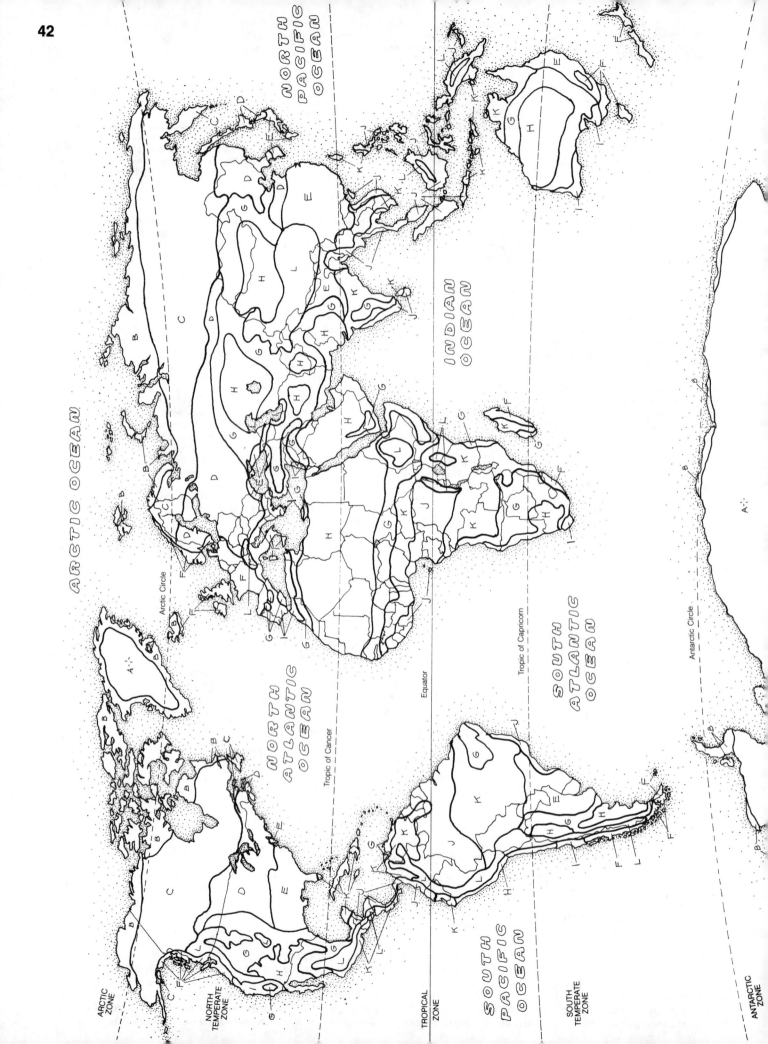

ARCTIC OCEAN

NORTH PACIFIC OCEAN

INDIAN OCEAN

NORTH ATLANTIC OCEAN

SOUTH ATLANTIC OCEAN

SOUTH PACIFIC OCEAN

Arctic Circle

Tropic of Cancer

Equator

Tropic of Capricorn

Antarctic Circle

ARCTIC ZONE

NORTH TEMPERATE ZONE

TROPICAL ZONE

SOUTH TEMPERATE ZONE

ANTARCTIC ZONE

# WORLD CLIMATE REGIONS

CN: (1) On the World Thematic Plates (42–47), color over the lightly drawn boundary lines of countries located within each area outlined by the darker lines. Use light colors so that the nations' boundaries will remain visible. Test your recall by trying to identify the nations in a particular region. Keep in mind that there are no clear divisions between one region and another. Climate changes are transitional and are not as abrupt as indicated on a map of this kind. (2) Do not color any lakes. (3) Do not color the names of the oceans. (4) You may wish to color one climate region on all continents first or you may prefer to color all the regions on one continent before going on to the next continent. (5) Note that the ice cap climates (A) are left uncolored Antarctica, shown at the bottom of the map, has been stretched far beyond its real size in order to indicate its relationship to the three continents of the southern hemisphere (see Plate 41 for a more accurate depiction). Antarctica has been left out of Plates 43–47.

## POLAR ZONES
ICE CAP A
POLAR (TUNDRA) B

## TEMPERATE ZONES
SUBPOLAR (SUBARCTIC) C
HUMID / CONTINENTAL D
HUMID / SUBTROPICAL E
MOIST / COASTAL F
STEPPE G
DESERT H
MEDITERRANEAN I

## TROPICAL ZONE
RAIN FOREST J
WET & DRY SAVANNA K
MOUNTAIN L

Climate is weather considered over a long period of time. Weather is the short-term condition of the atmosphere. The atmosphere is a layer of air 100 mi. (160 km) thick, surrounding the earth. Weather only occurs in the warmer and denser bottom 6 mi. (9.6 km) of the atmosphere. Air temperature, precipitation, wind velocity, air pressure, cloudiness, and humidity are the elements by which weather is measured.

The uneven heating of the earth's surface is the cause of all weather activity. These variations in the amount of radiation received from the sun largely depend on latitudinal position. In the tropics, the sun stays more or less overhead, creating eternal summer. The intense heat from the sun's direct rays causes ocean water to evaporate (warm air absorbs the most moisture); the tropics receive the heaviest rainfall. The amount of sunlight in the temperate zones varies according to season (see the diagram on Plate 41). The resulting fluctuation in heat creates the most variable weather on the planet. The sun's rays are least direct in the polar regions, and the result is almost constant cold weather.

Climate, particularly air temperature, is also affected by the proximity of large bodies of water. This is because water is colder than land during the summer and warmer than land during winter (water is slower to heat up than land in the summer, slower to cool down than land in the winter). This explains why regions such as the British Isles and western Europe have much milder climates (F) than other areas at similar latitudes. Because the interiors of Asia and North America are far from the influence of any ocean or sea, they tend to experience great temperature extremes, from hot summers to frigid winters. This type of climate is usually referred to as "continental."

Another major influence on climate is altitude. Mountain elevations (L) are generally much cooler, wetter, and windier than adjacent regions. Because the air is less dense at higher altitudes it contains fewer elements capable of retaining heat. This explains the presence of snow-capped mountains (Mt. Kilimanjaro in Africa) in the tropics. Mountains are generally wetter on their windward sides. Warm, moist air is swept upslope, cools down and releases its moisture before reaching the peaks.

Other factors that influence climate are wind patterns and ocean currents, which transfer heat and cold around the globe, and certain natural phenomena such as volcanic activity. Cataclysmic volcanic explosions and sustained periods of volcanic activity can spew enough dust into the atmosphere to block the sun's radiation, causing a drop of several degrees in global temperature. It is believed that the "Little Ice Age" of 1550–1880, the coldest period since the end of the most recent ice age (10,000 years ago), was the result of intense volcanic activity. Major ice ages, which have periodically covered much of the northern hemisphere with ice, are thought to follow changes in the earth's orbit around the sun.

Human beings are changing the world's climate through their activities. Scientists are growing concerned about an apparent rise in the earth's temperature. This "greenhouse effect" is caused by the increased production of carbon dioxide from the burning of fossil fuels (oil, coal, and natural gas). The excess carbon dioxide prevents the radiation of the sun's heat back into space. The widespread destruction of rain forests around the globe is intensifying the problem (plants absorb carbon dioxide). Plants may further be endangered by another byproduct of modern life, the thinning of the ozone layer in the atmosphere (see Plate 41).

The following classification of climates is based on temperature and precipitation, the two most important weather factors. They are treated separately on Plate 43. Most climates are generally found in one of the three basic earth zones: polar, temperate, and tropical. Some climates, such as desert (H) or mountain (L), are present in more than one zone.

POLAR ZONES. Ice cap is a below-freezing climate found in most of Greenland and all of Antarctica. The air is too cold to hold much moisture; there is only a little precipitation, in the form of light snow. Dryness and the absence of plants—almost nothing can grow on ice—give these regions a true desert status. Polar or tundra climate is always cold, though some regions experience brief, chilly summers of above-freezing temperatures. There is little precipitation. In the summer the upper inches of permafrost thaw. Cold air can hold little moisture, so evaporation is slow and the environment becomes wet and marshy. Wildflowers and low-growing plants make their appearance during this brief period.

TEMPERATE ZONES. Subpolar or subarctic climate is characterized by long, very cold winters and short, cool summers. Precipitation is light to moderate, and because of low evaporation, the flatter areas, with poor drainage, stay wet during the summer months. Coniferous trees cover parts of the landscape, and limited farming is possible. This is the climate of most of Canada and the northern USSR. Humid/continental climate is characterized by wide extremes in temperature (particularly in the interior regions of broad continents). Summers are normally mild but can become quite hot; winters are subject to periods of severe cold. Continental climate has moderate precipitation, most of it falling during the warm summer. Humid/subtropical climate has warm to hot summers and cool to cold winters and is subject to frequent cyclonic storms and highly variable weather. Rainfall is moderate, but summers can be very wet. These regions are found on the eastern sides of continents, in the lower latitudes of the temperate zone: the southeastern United States, southeastern South America, eastern China and Australia, and southern Japan. Moist/coastal, also called maritime or marine west coast climate, is moderately wet and is characterized by frequent cloudiness and light rain. Summers are milder and winters are less severe than in other regions within the same latitudes. This climate is generally found along the west coasts of continents in the upper latitudes of the temperate zone: western Europe, the British Isles, Canada, and the American northwest. In the southern hemisphere it is found in southern Chile, southeastern Africa and Australia, and New Zealand. Steppe is a dry climate with hot summers; it can have very cold winters, depending upon the latitude. There is a wide variation between day and evening temperatures. These transitional regions between deserts and the moister climates are often deprived of precipitation by adjacent mountain ranges. Steppes are found in large areas of the American West and Mexico, across the widest part of Africa south of the Sahara, in southcentral Asia, and encircling the western desert in Australia. Desert climates have very limited precipitation that is likely to fall in isolated downpours followed by long dry periods. The deserts of the higher temperate latitudes can experience very cold winters; those further to the south, such as the enormous Sahara, are hot all year long. A desert is a barren region with little or no rainfall. It is not necessarily sandy—only 20% of the Sahara is sandy. Some of the tropical deserts, such as those along the coasts of Peru, Chile, and Namibia, can go for many years without measurable rainfall. But since they are adjacent to the coast, these unusual deserts are often shrouded in fog. They are deprived of rain by cold ocean currents that cool the atmosphere, wringing moisture from the clouds before they can reach land. Mediterranean regions take their name from the climate in lands surrounding the Mediterranean Sea, which have very warm, dry summers and mild, wet winters. This climate is also found along parts of the west coast of continents in the lower temperate latitudes: central and southern California, central Chile, the Cape Town region of South Africa, and the southern coast of Australia. These regions of moderate temperatures, low humidity, and plentiful sunshine are generally viewed as very desirable places to live. Native trees and shrubs can survive long, dry periods.

TROPICAL ZONE. Rain forest temperatures are uniformly warm throughout the year. In the very humid rain forest climate, precipitation is heavy, varying from the Amazon Basin's almost daily afternoon downpours to the seasonal monsoons of Southeast Asia. Other wet equatorial areas are the Caribbean coast of Central America and the west coast of Africa. This hot and wet environment creates the lushest vegetation on earth. Wet and dry savanna climates are found in the tropics and are at times hotter than the rain forest. Rainfall is heavy only during the brief wet season. For the remainder of the year the savanna is dry. This climate characterizes large regions surrounding the rain forests of central Africa and the Amazon Basin in South America. Mountain climates can be found in any latitude. They are the result of cold or cool temperatures accompanying the high altitudes. Mountains are generally wetter and windier than surrounding environments, and many are permanently covered by snow and ice. Mountain climates are found in northwestern North America, central Mexico, the Andes in South America, the Tibetan Plateau and the mountains of Central Asia, and regions of Ethiopia and eastern Africa.

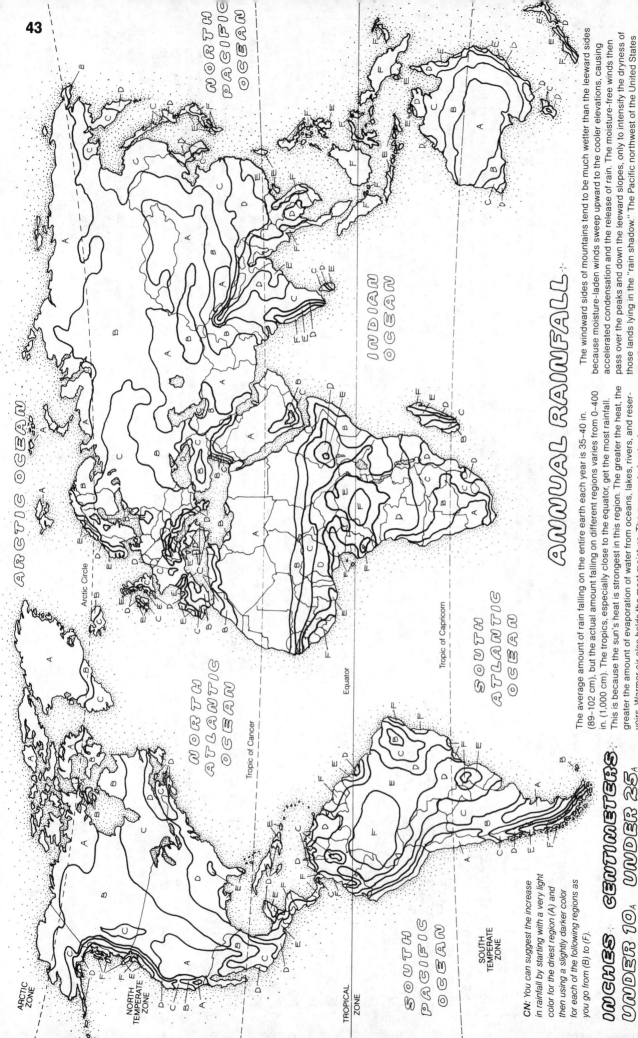

# 43

# ANNUAL RAINFALL

The average amount of rain falling on the entire earth each year is 35-40 in. (89-102 cm), but the actual amount falling on different regions varies from 0-400 in. (1,000 cm). The tropics, especially close to the equator, get the most rainfall. This is because the sun's heat is strongest in this region. The greater the heat, the greater the amount of evaporation of water from oceans, lakes, rivers, and reservoirs. Warmer air also holds the most moisture. The evaporated water becomes water vapor which rises in the warm air. The air cools at higher altitude and the water vapor condenses and forms clouds. Water droplets form and thousands of them combine to create each raindrop that is part of a rainfall.

Rainfall tends to be heaviest over or near large bodies of water. Rainfall is usually very light in the interiors of continents and in the ice-covered polar regions. It is heaviest where seasonal winds (monsoons) bring vast quantities of moisture from the sea. Monsoons contributed to the highest annual rainfall ever recorded: 1,042 in. (2,647 cm) in Cherrapunji, India, north of the Bangladesh border.

The windward sides of mountains tend to be much wetter than the leeward sides because moisture-laden winds sweep upward to the cooler elevations, causing accelerated condensation and the release of rain. The moisture-free winds then pass over the peaks and down the leeward slopes, only to intensify the dryness of those lands lying in the "rain shadow." The Pacific northwest of the United States and Canada shows how the presence of tall mountains can influence climate— very little rain can cross the massive ranges into the dry interior.

Unusual coastal deserts are found along the west coasts of Africa and South America where cold ocean currents, lying offshore, cool the moisture-laden clouds, causing rain to fall before the clouds can reach land.

Most rainfall is seasonal, but on the northwestern and eastern coasts of North America, in northern and western Europe, and on the eastern coast of South America, there is usually rainfall all year round. In some equatorial regions, such as the Amazon Basin, a day without rain is rare indeed.

*CN: You can suggest the increase in rainfall by starting with a very light color for the driest region (A) and then using a slightly darker color for each of the following regions as you go from (B) to (F).*

## INCHES      CENTIMETERS

| UNDER 10 | A | UNDER 25 | A |
|---|---|---|---|
| 10 – 20 | B | 25 – 50 | B |
| 20 – 40 | C | 50 – 100 | C |
| 40 – 60 | D | 100 – 150 | D |
| 60 – 80 | E | 150 – 200 | E |
| OVER 80 | F | OVER 200 | F |

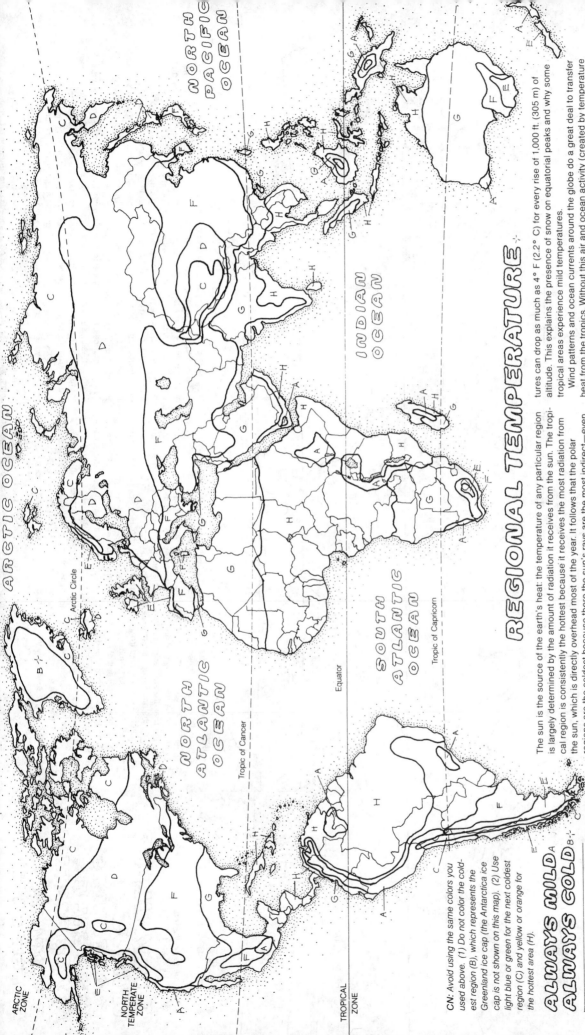

# REGIONAL TEMPERATURE

The sun is the source of the earth's heat: the temperature of any particular region is largely determined by the amount of radiation it receives from the sun. The tropical region is consistently the hottest because it receives the most radiation from the sun, which is directly overhead most of the year. It follows that the polar regions are the coldest because there the sun's rays are the most indirect—even in summer the polar sun stays close to the horizon. The temperate zones, in which the angle of the sun changes throughout the year (Plate 41), are the only regions to experience seasonal changes in temperature.

Other factors that influence temperature are proximity to large bodies of water, altitude, and prevailing wind conditions. Because water is much slower to heat up or cool down than land, the summer ocean still has some winter coolness and the winter ocean has not lost all of its summer warmth. The water temperature affects the ocean air, which exerts a moderating influence on the air above the coastal lands. Bodies of water are also heated or cooled by the flow of ocean currents. The coast of Norway is much warmer than regions at comparable latitudes in Canada or Asia, due to the presence of the North Atlantic Drift (Gulf Stream).

Altitude also influences temperature. As air rises, it becomes thinner and loses its chief heat-retaining constituents: water vapor and carbon dioxide. Tempera-

tures can drop as much as 4° F (2.2° C) for every rise of 1,000 ft. (305 m) of altitude. This explains the presence of snow on equatorial peaks and why some tropical areas experience mild temperatures.

Wind patterns and ocean currents around the globe do a great deal to transfer heat from the tropics. Without this air and ocean activity (created by temperature differences), the tropics would be much hotter and the rest of the globe much colder. As it is, temperature differences can still be immense: the highest temperature ever recorded was 136° F (58° C) in summer shade at Al Aziziyah, Libya, and the coldest was −128° F (−89° C) during winter in Antarctica.

For millions of years, the earth's overall temperature has remained relatively constant because of the "heat balance" between the amount of sun radiation absorbed by the earth and the radiation sent back to space. But carbon dioxide is building up in the earth's atmosphere because of the burning of fossil fuels, and a "greenhouse effect" is being created—radiation is being retained by the earth. Most scientists believe that a global warming trend has begun. Others insist that the earth is due for another ice age in which a temperature drop of only a few degrees will bring back the glaciers over much of the northern hemisphere. Ice ages have coincided with periodic variations in the earth's orbit around the sun.

*CN: Avoid using the same colors you used above. (1) Do not color the coldest region (B), which represents the Greenland ice cap (the Antarctica ice cap is not shown on this map). (2) Use light blue or green for the next coldest region (C) and yellow or orange for the hottest area (H).*

ALWAYS MILD_A
ALWAYS COLD_B

SUMMER: WINTER:
COOL_C    VERY COLD_D
MILD_D    COLD_D
MILD_E    COOL_E
HOT_F     COLD_F
VERY HOT_G  MILD_G

ALWAYS HOT_H

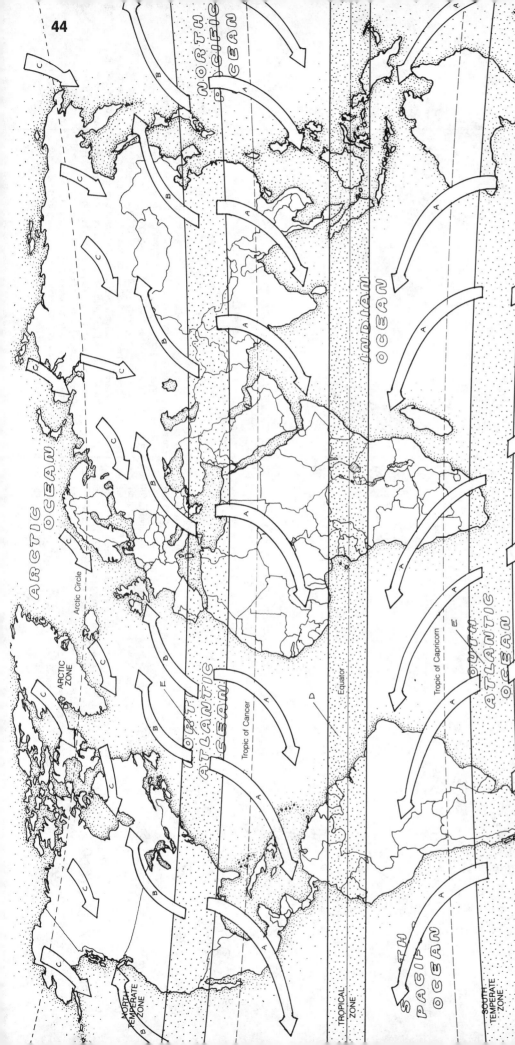

44

# PREVAILING WINDS :-

CN: (1) Use light colors on the three bands (D, and E twice) (2) Color the three types of winds (A–C). The polar easterlies in the southern hemisphere are not shown.

## TRADE WINDS A
## WESTERLIES B
## POLAR EASTERLIES C

## DOUDRUMS D
## HORSE LATITUDES E

Weather is generally the effect of large masses of air, warm or cold, wet or dry, moving around the planet. These masses originate as "source regions" where relatively stagnant air assumes a uniformity of moisture and temperature before beginning to spread its influence over great distances as a weather front. Air moves from high-pressure areas toward low-pressure areas, creating winds. Pressure gradients are caused by the uneven heating of the earth's atmosphere. Warm air around the equator rises from a low-pressure trough called the *doldrums* (which spans the equator between 5°N and 5°S latitude), and then flows north and south. When this air reaches a subtropical belt of high pressure in the *horse latitudes*, between 30° and 40° (N and S), it descends and becomes the *trade winds* blowing toward the equator and the *westerlies* blowing toward the polar regions. Because of the earth's rotation, these winds are deflected from a north/south movement and instead blow diagonally. Of the two "prevailing winds" (winds that blow most frequently in a particular region), the trades are the steadiest in force and direction. But the westerlies are generally stronger, especially in a 10° band of latitude just below the southern horse latitudes. Here, without any continental obstructions, winds can build up great force—both the winds and the latitudes are called the "roaring forties." The least predictable prevailing winds in subpolar regions, which mostly arise in subpolar regions,

south of the polar circles). Antarctica (not shown) has the most ferocious winds. The upward movement of warm air in the doldrums and the downward movement of cool air in the horse latitudes discourage any major horizontal movements, and some of the time these regions are without prevailing winds. The horse latitudes were supposedly so named because horses on sailing ships had to be killed when water ran short during periods of tropical stagnation. The name "doldrums" is self-evident. Tropical oceans spawn great windstorms that go by different names according to location: "hurricanes" in the Caribbean, "typhoons" in the southwest Pacific, and "cyclones" in the Indian Ocean.

Regional winds, often seasonal, can bring excessive moisture (the monsoons of Southeast Asia), searing heat (the sirocco from northern Africa to southern Europe); or unseasonable cold (from northern to southern Europe). Local winds occur in coastal areas: during the afternoon, cool ocean air rushes in to replace the heated air rising above the warm landscape. At night the situation is reversed, but winds heading toward the ocean are not as strong. In the interiors, winds are likely to ascend hillsides during the day and descend at night.

At much higher altitudes (5–12 mi., 8–20 km) the winds are dominated by westerlies; within them the jet streams flow at speeds up to 250 mph (400 kph).

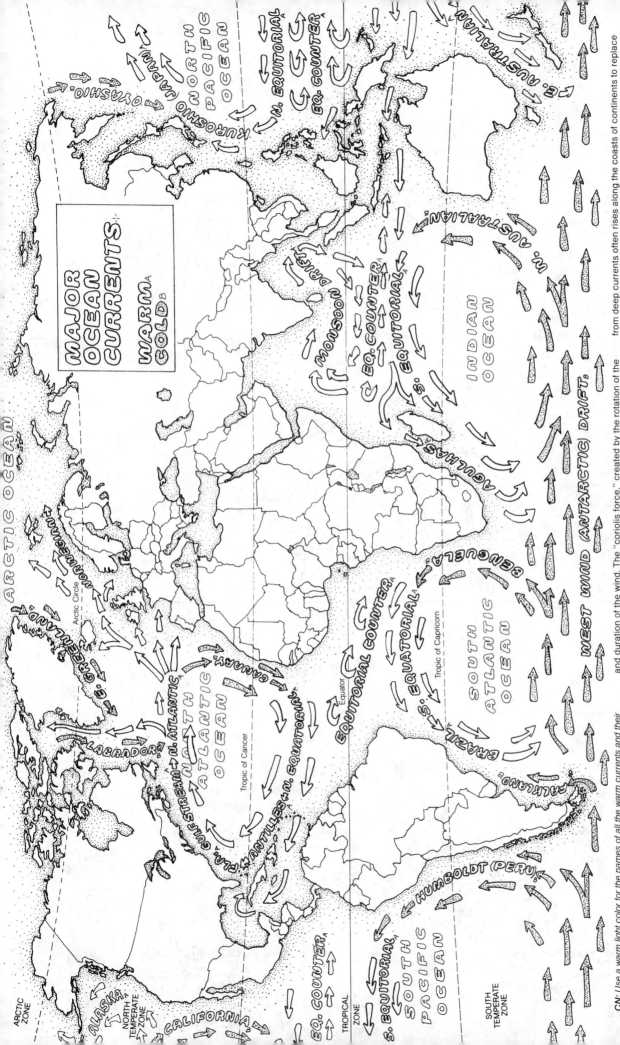

**MAJOR OCEAN CURRENTS**
WARM(A)  COLD(B)

ARCTIC ZONE

ARCTIC OCEAN

Arctic Circle

NORTH TEMPERATE ZONE

Tropic of Cancer

TROPICAL ZONE

Equator

Tropic of Capricorn

SOUTH TEMPERATE ZONE

KUROSHIO (JAPAN)

N. PACIFIC OCEAN

N. EQUATORIAL

EQ. COUNTER

W. AUSTRALIAN

N. AUSTRALIAN

MONSOON DRIFT

EQ. COUNTER(A)

S. EQUATORIAL(A)

INDIAN OCEAN

AGULHAS

E. AUSTRALIAN

WEST WIND (ANTARCTIC) DRIFT(B)

GREENLAND

S. GREENLAND

NORWEGIAN

LABRADOR(B)

GULF STREAM → N. ATLANTIC

NORTH ATLANTIC OCEAN

CANARY(B)

ANTILLES → N. EQUATORIAL

EQUATORIAL COUNTER(A)

S. EQUATORIAL(A)

BENGUELA(B)

SOUTH ATLANTIC OCEAN

BRAZIL

FALKLAND

ALASKA

NORTH TEMPERATE ZONE

CALIFORNIA

EQ. COUNTER(A)

S. EQUATORIAL

SOUTH PACIFIC OCEAN

HUMBOLDT (PERU)

**CN:** Use a warm light color for the names of all the warm currents and their directional arrows (A) and a cool light color for the cold currents and their dotted arrows (B). Because this world view shows only the edges of the Pacific Ocean, the important North Pacific current, which is an extension of the warm Kuroshio (Japan) current, is not shown.

Across the oceans stream broad "rivers" of warm and cool water. These "surface currents" are propelled mostly by prevailing winds; these winds create a general movement westward near the equator and an eastward movement in the higher latitudes (note the similarity to wind patterns shown on the upper map). Surface currents range in depth from up to 2,500 ft. (760m) and in speed from .5–5 mph (.8–9kph), depending upon the strength and duration of the wind. The "coriolis force," created by the rotation of the earth, exerts an influence on surface currents, causing them to veer to the right (clockwise) of the wind's direction in the northern hemisphere, to the left in the southern hemisphere. This force produces five major "gyres" (circular patterns), two in the northern oceans and three in the southern. Seasonal changes in wind directions similarly affect surface currents. The Monsoon Drift, shown below the Indian subcontinent, alternates in direction from summer to winter.

Beneath the surface currents, and moving at a much slower pace (about 1 mile per day), the cold, dense "deep currents" (not shown) move from the polar regions toward the warm and less dense waters of the equator. Water from deep currents often rises along the coasts of continents to replace warm surface currents that have been driven away from shore by the Coriolis force. These "upwellings" bring valuable nutrients to the surface waters and attract and support large fish populations.

Surface currents play an important role in moderating coastal climates by moving vast quantities of warm or cold water. The Gulf Stream, which is actually a system of different currents (North Equatorial, Antilles, Florida, Gulf Stream, and North Atlantic Drift), raises the average winter temperature of the British Isles and the west coast of Europe by 20° F (11° C). The southwestern coasts of continents in the southern hemisphere are cooled by currents spawned by the West Wind Drift, which encircles Antarctica.

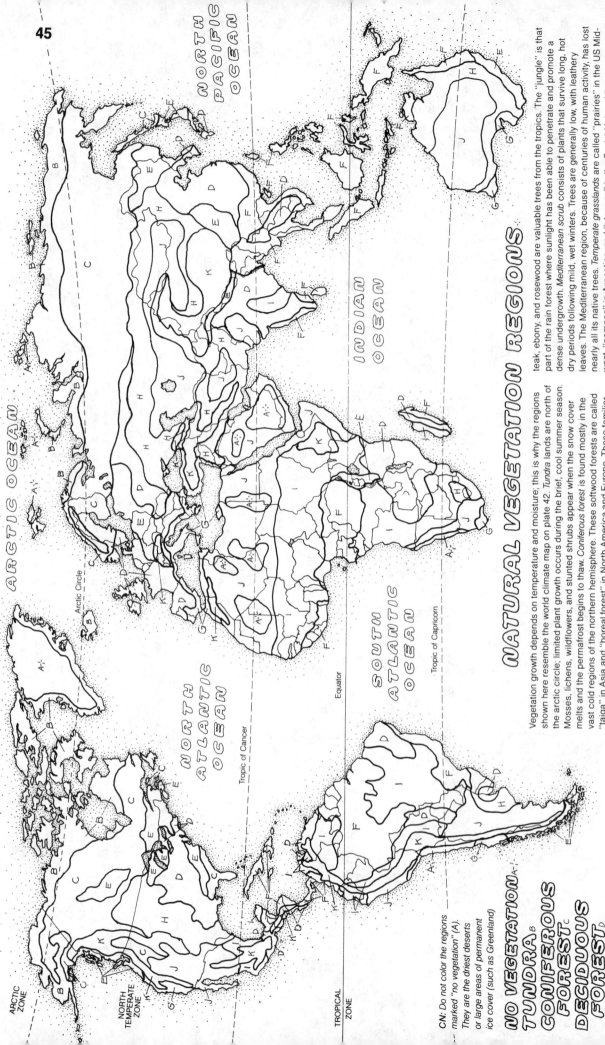

# 45

# NATURAL VEGETATION REGIONS

CN: Do not color the regions marked "no vegetation" (A). They are the driest deserts or large areas of permanent ice cover (such as Greenland).

NO VEGETATION_A
TUNDRA_B
CONIFEROUS FOREST_C
DECIDUOUS FOREST_D
MIXED CONIF & DECID_DE
TROPICAL RAIN FOREST_F
MEDITERRANEAN SCRUB_G
TEMPERATE GRASSLAND_H
TROPICAL GRASSLAND_I
DESERT SHRUB_J
MOUNTAIN_K

Vegetation growth depends on temperature and moisture; this is why the regions shown here resemble the world climate map on plate 42. Tundra lands are north of the arctic circle; limited plant growth occurs during the brief, cool summer season. Mosses, lichens, wildflowers, and stunted shrubs appear when the snow cover melts and the permafrost begins to thaw. Coniferous forest is found mostly in the vast cold regions of the northern hemisphere. These softwood forests are called "taiga" in Asia and "boreal forest" in North America and Europe. These familiar "Christmas trees" are evergreens with small needlelike leaves. Their conical shapes are able to withstand heavy snow buildups. The limited number of species include fir, spruce, larch, and pine. Deciduous forests of broadleaf hardwood trees once covered western Europe and the eastern half of the United States. Most of these trees in the temperate zone lose their leaves in the fall. Among the many varieties are oak, ash, beech, maple, hickory, and chestnut. Mixed coniferous and deciduous forests are transitional regions, containing both kinds of trees. Tropical rain forests contain an enormous number of species of broadleaf evergreen trees, jamming the rain-drenched equatorial regions. Tall trees with smooth, unbranched trunks form a solid canopy that darkens the forest floor. The trees are surrounded by light-seeking plants: "lianas" (thick climbing vines) and "epiphytes" (nonparasitic plants that grow on tree branches, such as mosses and orchids). Mahogany,

teak, ebony, and rosewood are valuable trees from the tropics. The "jungle" is that part of the rain forest where sunlight has been able to penetrate and promote a dense undergrowth. Mediterranean scrub consists of plants that survive long, hot dry periods following mild, wet winters. Trees are generally low, with leathery leaves. The Mediterranean region, because of centuries of human activity, has lost nearly all its native trees. Temperate grasslands are called "prairies" in the US Midwest, "pampas" in Argentina, and "steppes" in the Soviet Union. All are major wheat-producing areas. The soil is too dry to support the growth of trees. Tropical grasslands are called "savannas" in Africa (where they surround the rain forest and cover over a third of the continent); "llanos" in Venezuela; and "campos" in Brazil. Because tropical grasslands are subject to occasional seasons of heavy rain, they are likely to have scattered trees and the thickest grass. Deserts are not usually covered with sand dunes, nor are they totally devoid of vegetation. Most deserts have some drought-resistant shrubs or scrubby plants and spiny succulents. The cactus is a familiar sight in the American southwest. Most deserts are in the western portion of continents and lie between 25° and 30° latitude. Mountains usually have more than one vegetative zone, because temperature drops at higher elevations. A lofty peak in the tropics can emerge from a rain forest and

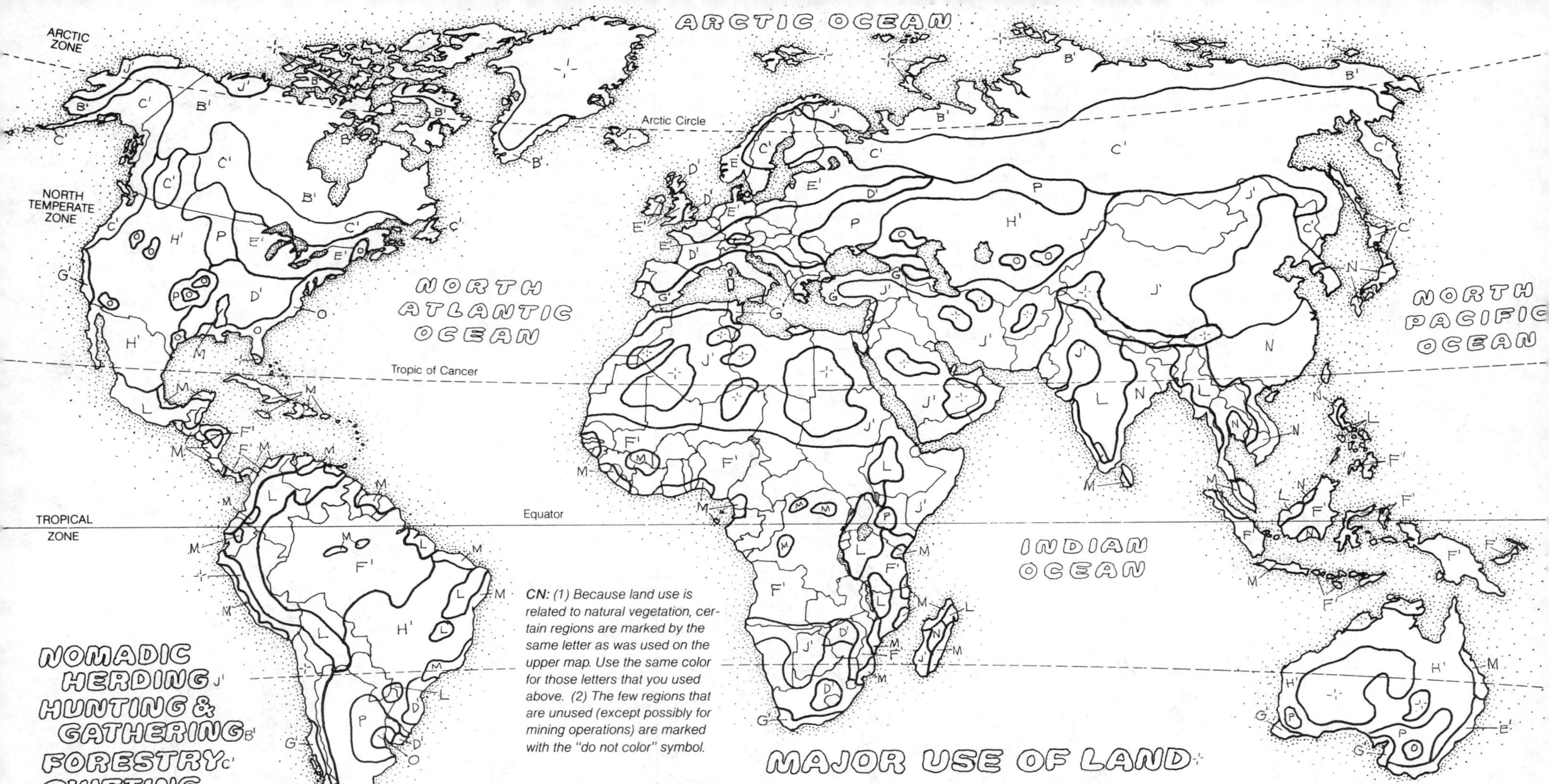

ARCTIC OCEAN

NORTH PACIFIC OCEAN

NORTH ATLANTIC OCEAN

INDIAN OCEAN

Arctic Circle

Tropic of Cancer

Equator

ARCTIC ZONE

NORTH TEMPERATE ZONE

TROPICAL ZONE

# MAJOR USE OF LAND

## Legend

NOMADIC HERDING J

HUNTING & GATHERING B

FORESTRY: C

SHIFTING CULTIVATION F'

SUBSISTENCE AGRICULTURE L

PLANTATIONS M

RICE PADDIES N

MIXED CROP & ANIMAL D'

MEDITERRANEAN G'

SPECIALIZED FARMING O

DAIRYING E'

LIVESTOCK RANCHING H

COMMERCIAL GRAIN P

CN: (1) Because land use is related to natural vegetation, certain regions are marked by the same letter as was used on the upper map. Use the same color for those letters that you used above. (2) The few regions that are unused (except possibly for mining operations) are marked with the "do not color" symbol.

Because they generally have the best weather and soil, the middle latitudes of the north temperate zone have produced the world's most advanced societies and the largest populations. Nomadic herding is still practiced in the deserts and semi-deserts of Africa and Asia. Because grass is scarce, tribes are always on the move. Hunting and gathering is still the way of life for the Eskimos in the arctic and the Bushmen in South Africa's Kalahari Desert. Like the nomads, hunter-gatherers are declining in number. Forestry—managing and replenishing timber wealth in the nations of the temperate zones—has become standard practice. This is not true in the tropics, where slow-growing hardwood forests are being rapidly destroyed. Shifting cultivation, also called "slash-and-burn," is an ancient agricultural practice. Tropical trees and plants are hacked down, allowed to dry, and then burned to the ground. The resulting farmland is marginally fertile and good only for two or three years. Subsistence agriculture methods barely meet the needs of farmers who use extensive human labor and very simple equipment to raise crops and cattle. Plantations are large tropical landholdings which specialize in growing and partially processing a single cash crop, such as coffee, bananas, sugar, tea, cacao, tobacco, or spices. With the help of slave labor, early plantations made great fortunes for the colonial owners. Rice paddies in the fields and hillside terraces of Southeast Asia are flooded by monsoon rains and irrigation networks to grow the

region's food staple. This ancient, highly intensive form of farming is practiced where land is scarce and maximum productivity is needed. A large investment is made in resources—labor, fertilizer, machinery, and water. Mixed crop and animal agriculture carries the least risk. It does not depend on the success of a single crop but instead relies on a wide variety for both human and animal consumption. North American and European farmers work their land "intensively"; Asian and Latin American peasants farm their land "extensively" (a lesser input of resources yields a lower productivity). Mediterannean farming specializes in drought-tolerant vines and trees (grapes, olives, and figs) that survive long, dry summers, and in grain crops that grow during mild, wet winters. Livestock is raised, and with the use of irrigation, citrus trees are grown. Specialized farming, also called "truck farming," is a highly intensive method, found mostly in temperate regions, that produces large amounts of fruit and vegetables for adjacent urban areas. Dairying in many areas takes place in highly automated "milk factories." In nations far from export markets, liquid milk is converted to processed milk, butter, or cheese. Livestock ranching depends upon plentiful land so that herds are able to find enough food even where grass is sparse. Commercial grain production is the most mechanized form of agriculture: giant farms are owned by absentee corporations whose employees are generally present only during planting and harvesting.

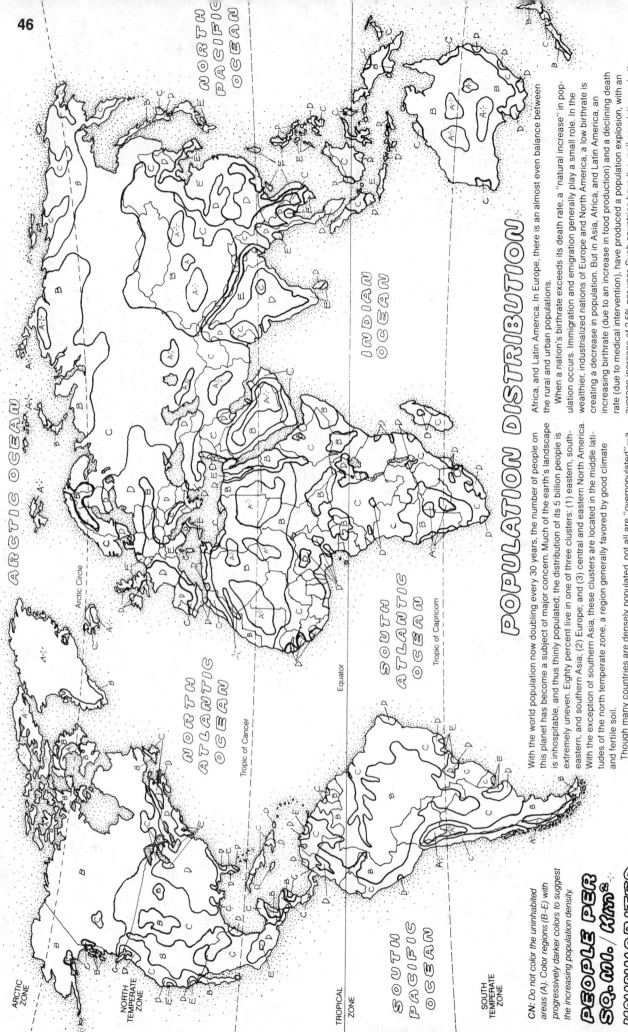

# 46

# POPULATION DISTRIBUTION

With the world population now doubling every 30 years, the number of people on this planet has become a subject of major concern. Much of the earth's landscape is inhospitable, and thus thinly populated; the distribution of its 5 billion people is extremely uneven. Eighty percent live in one of three clusters: (1) eastern, southeastern, and southern Asia; (2) Europe; and (3) central and eastern North America. With the exception of southern Asia, these clusters are located in the middle latitudes of the north temperate zone, a region generally favored by good climate and fertile soil.

Though many countries are densely populated, not all are "overpopulated"—a condition in which there are more people than an area can support. (For example, because of its high standard of living, the Netherlands, one of the most densely populated nations, is not considered overpopulated.) Uneven population distribution exists within nations themselves. The greatest concentrations are found in the job-producing urban areas ("urban" means having over 20,000 residents). Forty percent of the world's population now resides in or near cities, and the trend continues. In the United States, the figure is over 75 percent. In the past, people congregated only in food-producing areas and this is still true in much of Asia,

Africa, and Latin America. In Europe, there is an almost even balance between the rural and urban populations.

When a nation's birthrate exceeds its death rate, a "natural increase" in population occurs. Immigration and emigration generally play a small role. In the wealthier, industrialized nations of Europe and North America, a low birthrate is creating a decrease in population. But in Asia, Africa, and Latin America, an increasing birthrate (due to an increase in food production) and a declining death rate (due to medical intervention), have produced a population explosion, with an average increase of 2.5% per year. Such countries have "young" populations—half the people are under age 15. Farmers in the poorer nations depend upon children to perform free labor and provide old-age assistance. Parents in industrialized nations do not depend on child labor, and raising children can be very expensive (especially the costs of higher education). The elderly are also assisted by pensions and social security. So until the developing nations raise their standard of living—which ironically depends on a reduction in population growth—it is not likely that poor people will want to reduce the size of their families. China, with its rigid emphasis on the one-child family, is an exception.

**CN:** *Do not color the uninhabited areas (A). Color regions (B-E) with progressively darker colors to suggest the increasing population density.*

# PEOPLE PER SQ. MI. / KM²

| | |
|---|---|
| UNINHABITED | A |
| UNDER 2 / 1 | B |
| 2 - 60 / 1-25 | C |
| 60 - 250 / 25 - 100 | D |
| OVER 250 / 100 | E |

ARCTIC ZONE

NORTH TEMPERATE ZONE

TROPICAL ZONE

SOUTH TEMPERATE ZONE

ARCTIC OCEAN

NORTH PACIFIC OCEAN

NORTH ATLANTIC OCEAN

SOUTH ATLANTIC OCEAN

SOUTH PACIFIC OCEAN

INDIAN OCEAN

Arctic Circle

Tropic of Cancer

Equator

Tropic of Capricorn

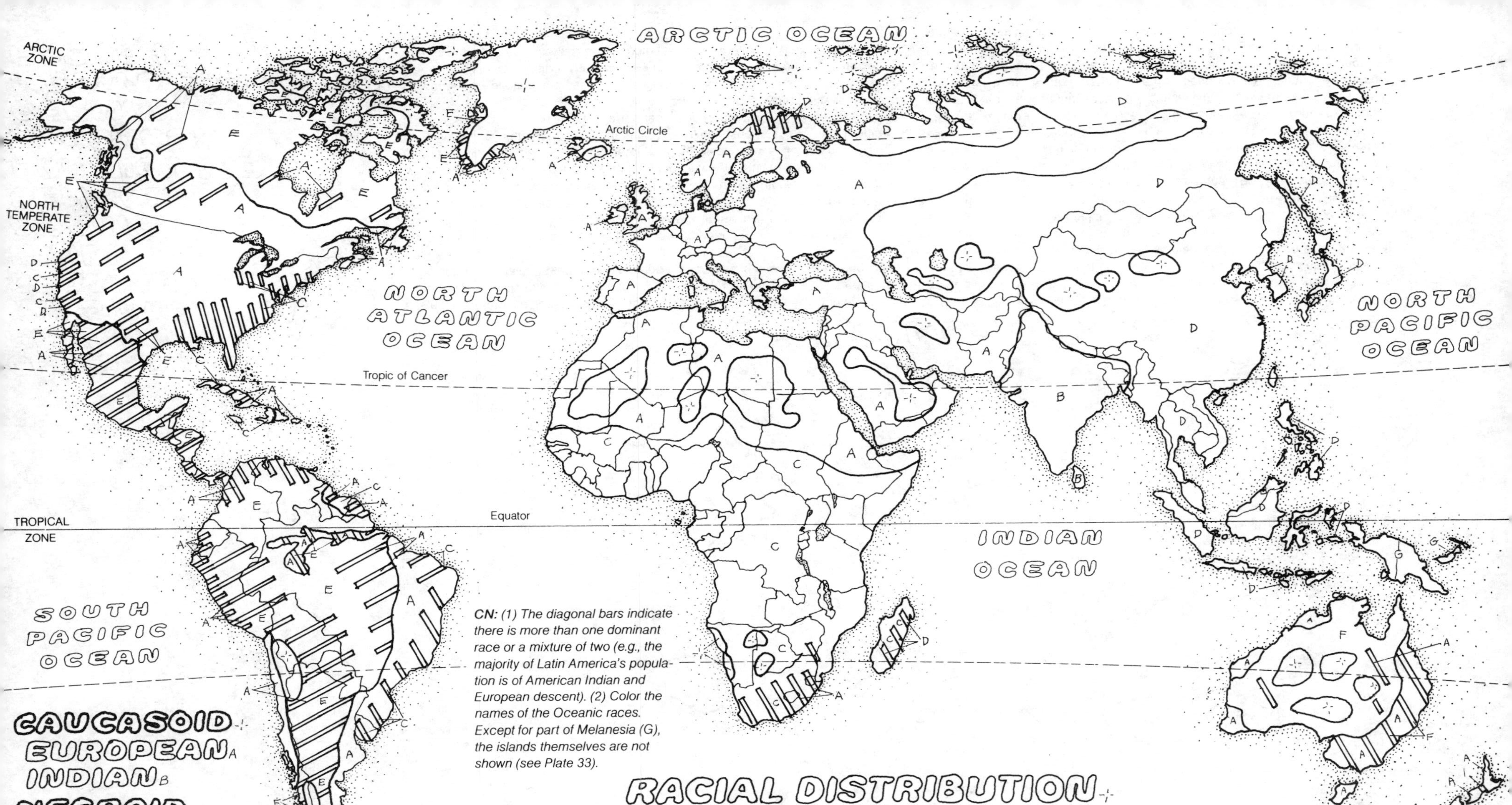

ARCTIC ZONE
NORTH TEMPERATE ZONE
TROPICAL ZONE

ARCTIC OCEAN

NORTH PACIFIC OCEAN

NORTH ATLANTIC OCEAN

INDIAN OCEAN

SOUTH PACIFIC OCEAN

Arctic Circle

Tropic of Cancer

Equator

## RACIAL DISTRIBUTION

CAUCASOID
EUROPEAN A
INDIAN B

NEGROID
AFRICAN C

MONGOLOID
ASIATIC D
AMERINDIAN E

OCEANIC
AUSTRALIAN F
MELANESIAN G
MICRONESIAN H
POLYNESIAN I

CN: (1) The diagonal bars indicate there is more than one dominant race or a mixture of two (e.g. the majority of Latin America's population is of American Indian and European descent). (2) Color the names of the Oceanic races. Except for part of Melanesia (G), the islands themselves are not shown (see Plate 33).

The term "race" refers to a group whose hereditary characteristics make its members resemble each other more than individuals of any other group. Race is a biological matter and has nothing to do with culture, nationality, religion, or behavioral characteristics. There is really only one human species, *Homo sapiens*. It is generally believed that all races evolved from the early inhabitants of eastern Africa. As groups emigrated to other parts of the world, their adaptation to different environments may have played a major role in determining variations in appearance. There is no single physical characteristic that is exclusive to any one race. The traditional classification—Caucasoid, Mongoloid, and Negroid—was based solely upon physical characteristics (skin color, hair texture; bone structure; shape of eyes, nose, and lips; etc.). A newer classification, based upon geographical origins, defines nine major racial groups (with hundreds of subgroups). For centuries, these races lived in virtual isolation because physical barriers surrounded their homelands. Racial "purity" began to disintegrate in the 15th century because of European exploration and colonization—half the people who have

some European ancestry do not live in Europe. Modern communication, commerce, and continued migrations are accelerating this process. The trend is toward race convergence: one race merging into another, creating groups with more variabilities.

There is great diversity within each race. The "white" *Caucasoid* race ranges from fair-skinned Scandinavians to dark-skinned North Africans and still darker Ethiopians. The dark-skinned residents of India are also considered Caucasoid. Blacks living in the Americas are descended from the *Negroid* races of sub-Saharan Africa. The *Mongoloid* race includes American Indians and Asiatics. The Americas were originally populated by red- and brown-skinned people who migrated from Asia across the Bering Strait, as did the Eskimos at a later date. The yellow-skinned Asiatics (Chinese, Japanese, Koreans, and Southeast Asians) share similar eye and facial characteristics. The *Oceanic* group includes the very dark-skinned Australian Aborigines and the progressively lighter Pacific Island races (Melanesians, Micronesians, and Polynesians: see Plate 33).

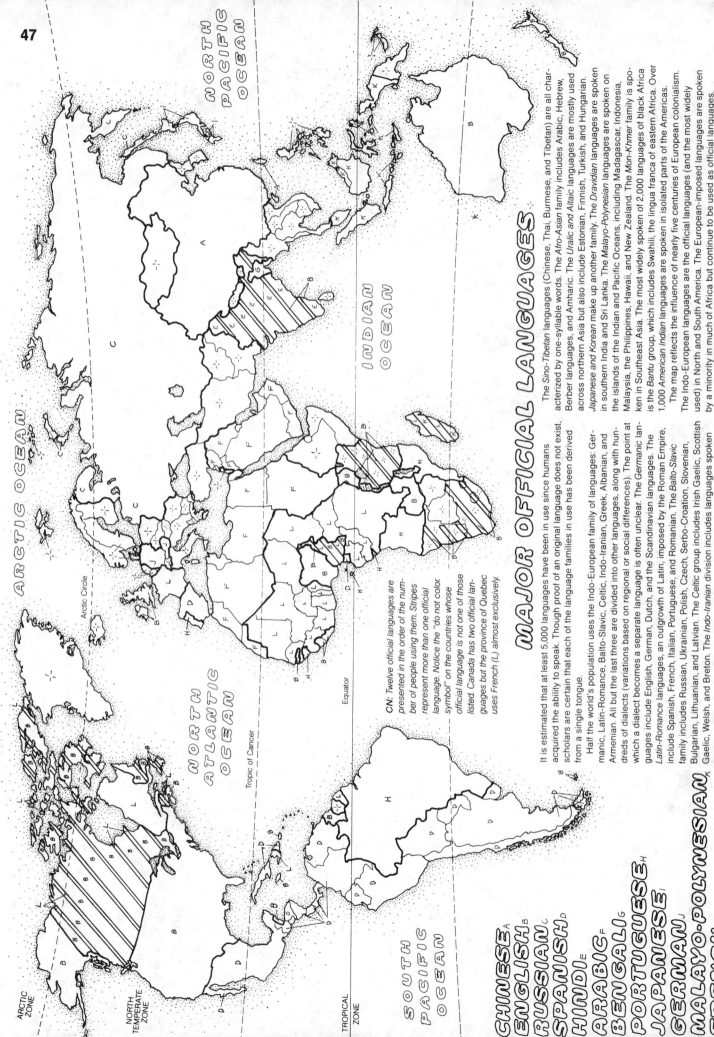

# 47

ARCTIC OCEAN

NORTH PACIFIC OCEAN

NORTH ATLANTIC OCEAN

INDIAN OCEAN

SOUTH PACIFIC OCEAN

ARCTIC ZONE

Arctic Circle

NORTH TEMPERATE ZONE

Tropic of Cancer

TROPICAL ZONE

Equator

CHINESE A
ENGLISH B
RUSSIAN C
SPANISH D
HINDI E
ARABIC F
BENGALI G
PORTUGUESE H
JAPANESE I
GERMAN J
MALAYO-POLYNESIAN K
FRENCH L

**CN:** Twelve official languages are presented in the order of the number of people using them. Stripes represent more than one official language. Notice the "do not color symbol" on the countries whose official language is not one of those listed. Canada has two official languages but the province of Quebec uses French (L) almost exclusively.

# MAJOR OFFICIAL LANGUAGES

It is estimated that at least 5,000 languages have been in use since humans acquired the ability to speak. Though proof of an original language does not exist, scholars are certain that each of the language families in use has been derived from a single tongue.

Half the world's population uses the Indo-European family of languages: Germanic, Latin-Romance, Balto-Slavic, Celtic, Indo-Iranian, Greek, Albanian, and Armenian. All but the last three are divided into other languages, along with hundreds of dialects (variations based on regional or social differences). The point at which a dialect becomes a separate language is often unclear. The *Germanic* languages include English, German, Dutch, and the Scandinavian languages. The *Latin-Romance* languages, an outgrowth of Latin, imposed by the Roman Empire, include Spanish, French, Italian, Portuguese, and Romanian. The *Balto-Slavic* family includes Russian, Ukrainian, Polish, Czech, Serbo-Croation, Slovenian, Bulgarian, Lithuanian, and Latvian. The *Celtic* group includes Irish Gaelic, Scottish Gaelic, Welsh, and Breton. The *Indo-Iranian* division includes languages spoken across southern Asia: Farsi (Iran), Pashto (Afghanistan), Urdu (Pakistan), Hindi (India), and Bengali (Bangladesh).

The *Sino-Tibetan* languages (Chinese, Thai, Burmese, and Tibetan) are all characterized by one-syllable words. The *Afro-Asian* family includes Arabic, Hebrew, Berber languages, and Amharic. The *Uralic and Altaic* languages are mostly used across northern Asia but also include Estonian, Finnish, Turkish, and Hungarian. *Japanese and Korean* make up another family. The *Malayo-Polynesian* languages are spoken in southern India and Sri Lanka. The *Malayo-Polynesian* languages are spoken on the islands of the Indian and Pacific Oceans, including Madagascar, Indonesia, Malaysia, the Philippines, Hawaii, and New Zealand. The *Mon-Khmer* family is spoken in Southeast Asia. The most widely spoken of 2,000 languages of black Africa is the *Bantu* group, which includes Swahili, the lingua franca of eastern Africa. Over 1,000 *American Indian* languages are spoken in isolated parts of the Americas.

The map reflects the influence of nearly five centuries of European colonialism. The Indo-European languages are the official languages (and the most widely used) in North and South America. The European-imposed languages are spoken by a minority in much of Africa but continue to be used as official languages. (There is no common tongue among the many ethnic groups in many countries and the ruling and business classes have been educated in European languages.)

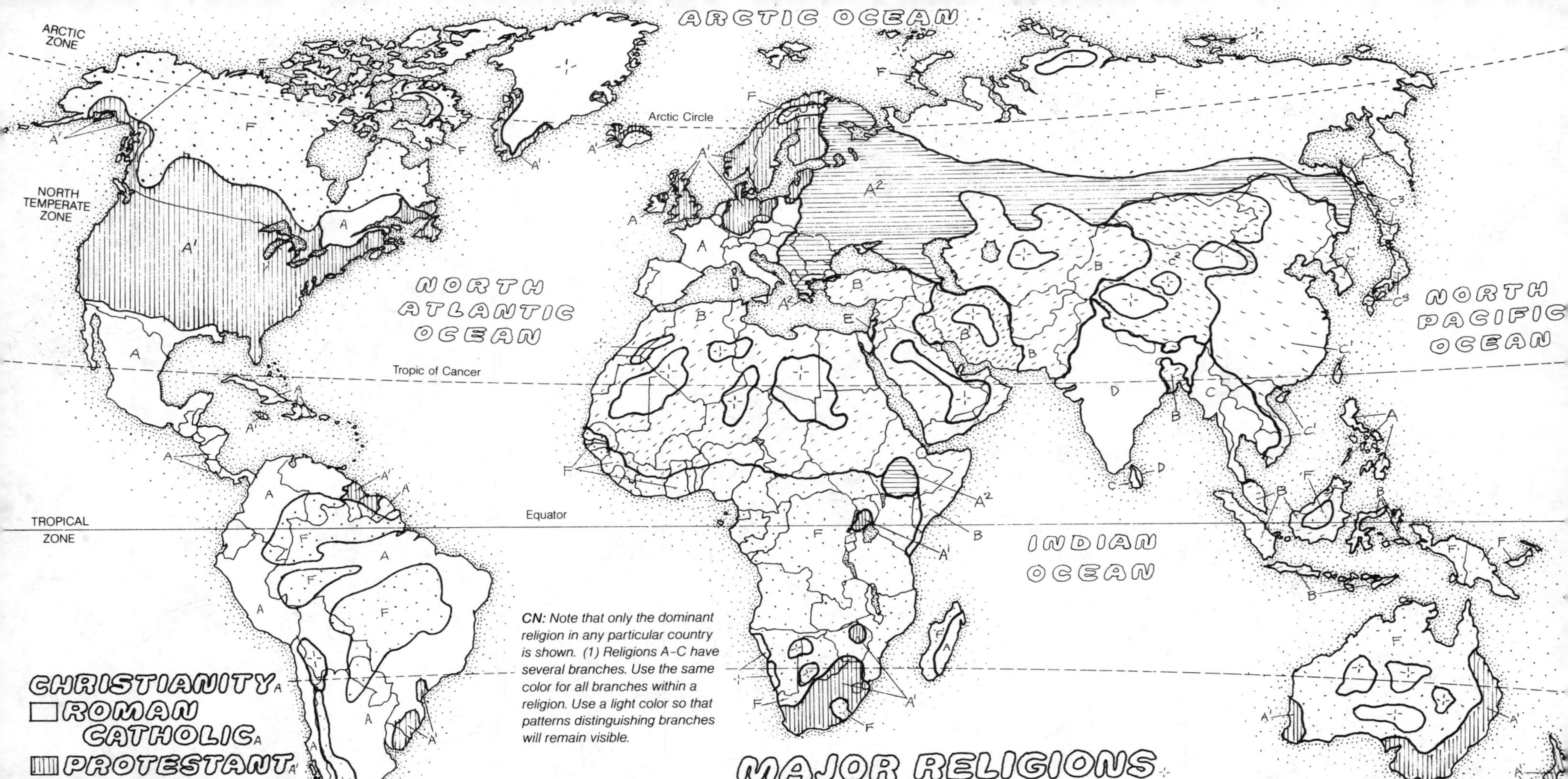

ARCTIC OCEAN

NORTH PACIFIC OCEAN

NORTH ATLANTIC OCEAN

INDIAN OCEAN

Arctic Circle

Tropic of Cancer

Equator

ARCTIC ZONE

NORTH TEMPERATE ZONE

TROPICAL ZONE

*CN: Note that only the dominant religion in any particular country is shown. (1) Religions A–C have several branches. Use the same color for all branches within a religion. Use a light color so that patterns distinguishing branches will remain visible.*

**CHRISTIANITY** A
☐ **ROMAN CATHOLIC** A
▤ **PROTESTANT** A¹
▥ **EASTERN** A²
**ISLAM** B
☐ **SUNNI** B
▨ **SHIITE** B¹
**BUDDHISM** ☐ C
☐ **CONFUCIANIST** C¹
☐ **TAOIST** C¹
☐ **LAMAIST** C²
☐ **SHINTO** C³
**HINDUISM** D
**JUDAISM** E
**ANIMISM** F

## MAJOR RELIGIONS

*Judaism* is the oldest of the monotheistic religions. The tribal leader Abraham was believed to have received land from God about 3,400 years ago. Jews, Christians, and Muslims believe that Moses was a prophet who was given God's law. The Torah (sacred writings of the word of God) is the Old Testament of the Christian Bible. The 17 million followers of Judaism are mainly ethnic Jews. *Christianity*, with over 1.5 billion adherents, is the largest religion. It was founded 2,000 years ago by a Jew, Jesus Christ, whom Christians regard as their Savior and Messiah. In 1054 Christianity split into the *Roman Catholic* Church, led by the Pope of Rome, and the *Eastern (Orthodox)* group of independent churches who followed early Christian tenets. Five hundred years afterwards, the Protestant Reformation split the Roman Catholic Church; new denominations spread across the Germanic-speaking nations of northern Europe and later to the United States and most of Canada. Protestants objected to the practices of the papacy and the priesthood and sought revelation solely from the Bible. There are twice as many Catholics as there are Protestants. *Islam*, the third major religion born in the Middle East, was founded about 1400 years ago by Muhammed, who regarded himself as the last great prophet, following Moses and Christ. Five times each day, over 800 million Muslims kneel in prayer towards Mecca, the holy city in Saudi Arabia. Their God is Allah and their scripture is the Koran. Most Muslims are *Sunnis*. The fundamentalist *Shiite* minority are concentrated in Iran and Iraq. Islam is rapidly growing and has been more successful than Christianity in converting black animists in Africa. *Animism* is the worship of spirits believed to inhabit all natural forms and forces. *Hinduism*, the dominant religion of India and Nepal, involves the worship of thousands of deities. It has no clergy and no formal scripture other than religious writings called the Vedas. Hinduism has no single founder and has undergone many changes in its 5,000-year existence. Over half a billion believers seek purification and an end to the cycle of rebirths resulting from one's "Karma" (deeds in a previous life). Reincarnation takes many forms; hence, Hindus revere all manner of life. *Buddhism*, an offshoot of Hinduism, was founded about 1,500 years ago by Gautama Buddha. The Buddha ("the enlightened one") advocated moderation as a way to overcome desire and ambition and thus reach a state of contentment ("nirvana"), ending the cycle of painful reincarnations. Individual meditation is more important to nearly a half billion Buddhists than participation in ritualistic ceremonies. Traditional Buddhism is followed only in Southeast Asia. Elsewhere it has been modified by regional folk religions: in China by the *Confucian* respect for elders, ancestry, and authority and the *Taoist* philosophy of passivity and simplicity; in Tibet and Mongolia by *Lamaistic* monasticism and demonology; and in Japan by the spiritualism and rituals of *Shintoism* and the contemplative nature of Zen.

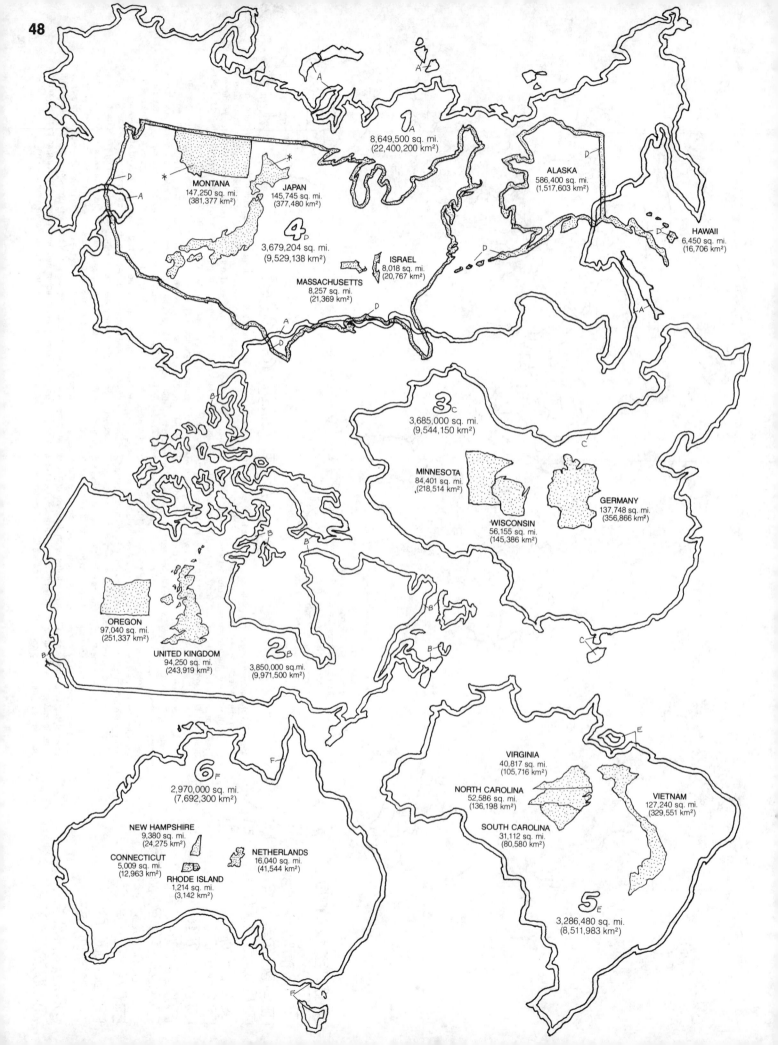

48

**1**<sub>A</sub>
8,649,500 sq. mi.
(22,400,200 km²)

ALASKA
586,400 sq. mi.
(1,517,603 km²)

MONTANA
147,250 sq. mi.
(381,377 km²)

JAPAN
145,745 sq. mi.
(377,480 km²)

HAWAII
6,450 sq. mi.
(16,706 km²)

**4**<sub>D</sub>
3,679,204 sq. mi.
(9,529,138 km²)

ISRAEL
8,018 sq. mi.
(20,767 km²)

MASSACHUSETTS
8,257 sq. mi.
(21,369 km²)

**3**<sub>C</sub>
3,685,000 sq. mi.
(9,544,150 km²)

MINNESOTA
84,401 sq. mi.
(218,514 km²)

WISCONSIN
56,155 sq. mi.
(145,386 km²)

GERMANY
137,748 sq. mi.
(356,866 km²)

OREGON
97,040 sq. mi.
(251,337 km²)

UNITED KINGDOM
94,250 sq. mi.
(243,919 km²)

**2**<sub>B</sub>
3,850,000 sq.mi.
(9,971,500 km²)

VIRGINIA
40,817 sq. mi.
(105,716 km²)

NORTH CAROLINA
52,586 sq. mi.
(136,198 km²)

VIETNAM
127,240 sq. mi.
(329,551 km²)

SOUTH CAROLINA
31,112 sq. mi.
(80,580 km²)

**6**<sub>F</sub>
2,970,000 sq. mi.
(7,692,300 km²)

NEW HAMPSHIRE
9,380 sq. mi.
(24,275 km²)

CONNECTICUT
5,009 sq. mi.
(12,963 km²)

NETHERLANDS
16,040 sq. mi.
(41,544 km²)

RHODE ISLAND
1,214 sq. mi.
(3,142 km²)

**5**<sub>E</sub>
3,286,480 sq. mi.
(8,511,983 km²)

# COMPARATIVE SIZES

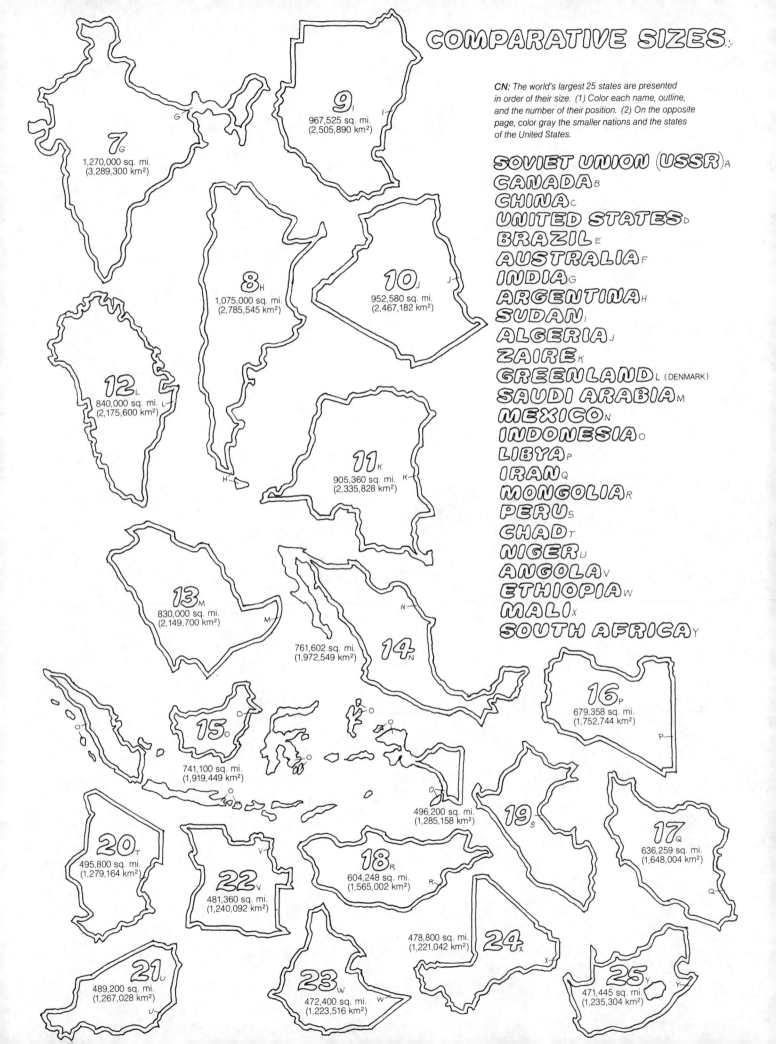

**CN:** The world's largest 25 states are presented in order of their size. (1) Color each name, outline, and the number of their position. (2) On the opposite page, color gray the smaller nations and the states of the United States.

SOVIET UNION (USSR)A
CANADA B
CHINA C
UNITED STATES D
BRAZIL E
AUSTRALIA F
INDIA G
ARGENTINA H
SUDAN I
ALGERIA J
ZAIRE K
GREENLAND L (DENMARK)
SAUDI ARABIA M
MEXICO N
INDONESIA O
LIBYA P
IRAN Q
MONGOLIA R
PERU S
CHAD T
NIGER U
ANGOLA V
ETHIOPIA W
MALI X
SOUTH AFRICA Y

7 G
1,270,000 sq. mi.
(3,289,300 km²)

9 I
967,525 sq. mi.
(2,505,890 km²)

8 H
1,075,000 sq. mi.
(2,785,545 km²)

10 J
952,580 sq. mi.
(2,467,182 km²)

12 L
840,000 sq. mi.
(2,175,600 km²)

11 K
905,360 sq. mi.
(2,335,828 km²)

13 M
830,000 sq. mi.
(2,149,700 km²)

761,602 sq. mi.
(1,972,549 km²)

14 N

16 P
679,358 sq. mi.
(1,752,744 km²)

15 O
741,100 sq. mi.
(1,919,449 km²)

496,200 sq. mi.
(1,285,158 km²)

19 S

17 Q
636,259 sq. mi.
(1,648,004 km²)

20 T
495,800 sq. mi.
(1,279,164 km²)

22 V
481,360 sq. mi.
(1,240,092 km²)

18 R
604,248 sq. mi.
(1,565,002 km²)

478,800 sq. mi.
(1,221,042 km²)

24 X

21 U
489,200 sq. mi.
(1,267,028 km²)

23 W
472,400 sq. mi.
(1,223,516 km²)

25 Y
471,445 sq. mi.
(1,235,304 km²)

*Where more than one page number is shown, the bold number indicates the principal reference: the page on which the subject is either colored, given prominent display, or written about. The other references generally show the subject in the context of other principal environments.*

# INDEX

# INDEX

INDEX

# INDEX